AF556240

Textbook of

Plant Propagation and Nursery Management

R.R. Sharma
Senior Scientist (Fruit Science)
Division of Post Harvest Technology
I.A.R.I., New Delhi - 12

Hare Krishna
Senior Scientist (Fruit Science)
CIAH, Beechwal,Bikaner-334 006
Rajasthan

Published by

Khushnuma Complex Basement
7, Meerabai Marg (Behind Jawahar Bhawan)
Lucknow 226 001 U.P. (INDIA)
Tel. : 91-522-2209542, 2209543, 2209544, 2209545
Fax : 0522-4045308
E-Mail : ibdco@airtelmail.in

First Edition 2013

ISBN 978-81-8189-198-3 (HB)

ISBN 978-81-8189-199-0 (PB)

Composed & Designed at :
Panacea Computers
3rd Floor, Agrawal Sabha Bhawan
Subhash Mohal, Sadar Cantt. Lucknow-226 002
Phone : 0522-2483312, 9335927082, 9452295008
E-mail : prasgupt@rediffmail.com

Printed at:
Salasar Imaging Systems
C-7/5, Lawrence Road Industrial Area
Delhi - 110 035
Tel. : 011-27185653, 9810064311

Dedicated

to:

Our most respected parents and teachers

डा. एस. अय्यप्पन

सचिव एवं महानिदेशक

Dr. S. AYYAPPAN

SECRETARY & DIRECTOR GENERAL

भारत सरकार

कृषि अनुसंधान और शिक्षा विभाग एवं

भारतीय कृषि अनुसंधान परिषद्

कृषि मंत्रालय, कृषि भवन, नई दिल्ली 110 114

GOVERNMENT OF INDIA
DEPARTMENT OF AGRICULTURAL RESEARCH & EDUCATION
AND
INDIAN COUNCIL OF AGRICULTURAL RESEARCH
MINISTRY OF AGRICULTURAL, KRISHI BHAVAN, NEW DELHI 110 114
Tel.: 23382629; 23386711 Fax: 91-11-23384773
E-mail : dg.icar@nic.in

FOREWORD

India is witnessing a rapid growth of horticulture and presently its share in agricultural GDP is about 29 percent. The sector also contributes nearly 37% of the total export of agricultural commodities from about 13 percent of area. Horticultural sector has long been hailed for improving the productivity of land, generating employment, improving economic conditions of the farmers and entrepreneurs, enriching exports and above all improving the nutritional security.

The productivity of major horticultural crops in the country is low. Several factors contribute to low productivity but lack of quality planting material is considered as the most important factor. Further, in our country, there are about 50 thousand public and private nurseries but still there is a paucity of quality planting materials. Plant propagation and nursery management course is taught to UG and PG students in every agricultural university. Thus, there is urgent need for a textbook which covers all content of the syllabus on plant propagation and nursery management. Therefore, to achieve the desired growth in horticulture sector, it is very important to ensure the supply of the quality planting materials.

I congratulate the authors, Dr. R.R. Sharma and Dr. Hare Krishna of the book entitled 'Textbook of Plant Propagation and Nursery Management' who have brought out a textbook for the undergraduate and post-graduate students in the present form. The authors have put considerable efforts to include each and every aspect of plant propagation and nursery management in this book, which has been written in a very lucid language with ample numbers of figures for easy understanding of the students, nurserymen and alike.

It is hoped that this book will be very useful for teachers, scientists, researchers, UG and PG students, extension personnel, and horticultural professionals.

(S. Ayyappan)

PREFACE

"सुबीजम् सुक्षेत्रे जायते सम्पद्यते"

ऋग्वेद

'A good seed in a good soil yields abundant'.

Rigveda

The importance of a good seed in agricultural production can well be understood by acknowledgements made in various ancient scriptures. Botanically, seed is a result of fertilization between male and female gametes; however, in horticulture a seed refers to any plant propagules, which is used for further multiplication. Presently, horticulture contributes 29% of GDP from about 13.08% of area. The developments made in previous decades have now placed the country as the 2nd largest producers of fruits and vegetables. Besides, we are the leading producers of many spices and plantation crops. The total annual horticultural production has been estimated to be about 230 MT, presently. However, keeping in view the future demand for horticultural produce, which has been estimated to be 300 MT at the end of 2011-2012 and 360 MT by the year 2020 and one of the lowest per capita availability of horticultural produce in world, revolutionary steps need to be adopted. Since, there is a limited scope for further area expansion under horticultural crops; therefore, there is a need to improve the return out put from the existing covered areas. One such measure could be the replacement of old senile orchards with improved fruit cultivars. Likewise, for vegetables and other annual crops, which are propagated through sexual means, there is an urgency to have sufficient quantity of seeds of high yielding improved cultivars. Further, for expanding area in waste and degraded land, a large quantity of planting material will be required. But, lack of quality planting materials to cater this huge demand mars the prospect of mitigating the challenges faced by horticultural industry. In addition, the lack of knowledge of basic principles and practices of propagation among nurserymen and growers aid in aggravating this problem further.

At present, a large number of agricultural students are opting for horticultural science as their major subject, in view of prospects of employment and entrepreneurship. Plant propagation and nursery management is an important aspect of horticultural production and it constitutes a full course in graduation. The present book is written keeping in mind a uniform syllabus prepared by ICAR for such undergraduate students. Besides dealing with principles and practices of propagation and nursery management, this book contains a model question bank comprising short notes, multiple choice questions, fill in the blanks and true and false statements. This would help them writing any competitive exams such as ICAR's JRF for post graduation. This book will also serve the purpose of nurserymen, amateur gardeners and even post graduate students due to its well fabricated content.

This book could not have come in its present form without the helps received from various colleagues and students in different stages during the preparation of the manuscript and for which they are profusely thanked.

We also acknowledge our family members who were ever there to extend unconditional support during the preparation of this book and we thank all of them from the bottom of our hearts.

RR Sharma | Hare Krishna

About the Authors

Dr. R.R. Sharma, Born on March 14, 1962 in Village Jajri, Tehsil Barsar, Distt. Hamirpur (Himachal Pradesh), Dr. Sharma completed his B.Sc. (Agri.) from HPKV, Palampur during 1983. He joined IARI, New Delhi-12 as Technical Assistant during 1985 and completed his Post Graduation from IARI, New Delhi during 1994 and secured top position in the Discipline of Horticulture. He qualified ARS-1995 and joined as Scientist in the Division of Fruits and Horticultural Technology, IARI, New Delhi in 1997, and completed his doctorate in 2002. During his scientific career, he has served several research projects and has been associated with release and assessment of mango hybrids like Pusa Arunima, Pusa Surya, Pusa Lalima, Pusa Shresth, Pusa Pratibha, Pusa Pitamber etc., and several Fruit Production technologies. He is a prolific writer and a scientist of international repute, as he has published 30 research articles in International journals, 50 in National journals, 160 popular articles in different magazines, and 25 book chapters. He has also authored 7 books on different aspects of fruit production. He has delivered several radio and T.V talks on different aspects of fruit and post harvest management of horticultural crops. He has received several honours and awards, of which Dr. R.N. Singh award of IARI, New Delhi and Dr Rajinder Prasad award of ICAR, New Delhi are worth mentioning. He has served as International Mango Registrar in International Society for Horticultural Science for about 5 years (1999-2004). He has been honoured by International journals like Scientia Horticulturae (The Netherlands), International Journal of Fruit Science (U.K.) and Stewart Postharvest Reviews (UK) by including him as a Member, Editorial Board. Recently, he has been awarded with Dr. Rajinder Prasad Award - 2010 by the ICAR.

Dr. Hare Krishna, Born on June 28, 1981, Dr Hare Krishna graduated from the Institute of Agricultural Sciences, B.H.U., Varanasi and completed his M. Sc. and Ph. D. from Indian Agricultural Research Institute, New Delhi. He has got an outstanding academic career. With his experience and vision, he contributed significantly to address the major obstacles of commercial micropropagation. He identified suitable mycorrhizal species for the bio-hardening of tissue cultured grape varieties released from IARI, New Delhi. Based on this work a protocol for bio-hardening of tissue culture-raised plantlets was developed. Furthermore, this study also earned him the prestigious IARI Gold Medal. He has also developed insights in to the problems of mango tissue culture particularly explant browning, which is a most important hurdle in the way of successful commercialization of mango micropropagation. He was invited for the creation of a Scirus Topic page on Mango Tissue Culture for the scientific community run by Elsevier, The Netherlands. He began his professional career by joining Agriculture Research Service in 2006 at Vivekanand Institute of Hill Agriculture, Almora. Later he was transferred to Central Institute of Temperate Horticulture-Regional Station, Mukteshwar. And since then, he has been dedicated to the field of temperate fruit production. During his initial years of career, so far he has published research papers in 08 international and 09 national journals. Besides, presently he is reviewer of several international journals such as Scientia Horticulturae, Crop Protection, Journal of American Society for Horticultural Science etc.

Contents

1 Introduction

Ever since the dawn of civilization, propagation of agriculturally important plants has been one of the primary occupations of mankind as a measure to secure livelihood. The pre-historic man subsisted partly by collecting seeds of wild edible plants, mostly grasses and various pulses. In due course of time, they learnt cultivation of the cereal crops for their survival and with advancement; many horticultural plants were also brought under cultivation. Though, fleshy fruits and dry nut fruits were the inevitable part of their diet, and it was only several thousands of years later that domestication of woody perennial fruit crops took off. Unlike annual cereals and pulses, which are easily propagated through seeds and come true-to-type due to self pollination, most of the fruit crops are cross pollinated, highly heterozygous in nature and could not be perpetuated through sexual means. As a result, the progeny of woody plants grown from seed tend to exhibit a relatively high amount of genetic variations.

Later, progress was made on the selection of the desirable types from the wild population (tomato, barley, rice etc.), natural hybrids (strawberry) and natural population (cole crops). But this progress would not have any significance, if man could not have any method to maintain them. Most of the selected plants would have lost or could have attained undesirable forms if they were not propagated under strict environmental control. This might have forced the mankind to think about alternative means. The initial attempts might have started; of course serendipitously, with multiplication of easy-to-root fruit plants such as grape, fig, olive and pomegranate through cutting and propagation by division of rhizome or separation of offshoots and runners as in banana, date palm and strawberry, respectively. The propagation of difficult-to-root plants, which include most of temperate fruits, was started as early as in first millennium BCE (Before the Christian Era). The grafting technique might have been evolved by encountering natural grafts between two adjacent intact plants or different branches, which are not very uncommon in nature. Many new methods of budding and grafting have been standardized with the evolution of civilization and development of social skills, which in turn reinforced mankind to attain food security.

The practice of vegetative propagation of fruit trees dates back to ancient times. China had been using grafting techniques since pre-classical times. The Greeks and the Romans adopted this strategy and spread these methods all over the Europe. The technique of grafting was discussed in detail by the European historic writers like Aristotle and Theophrastus and it became very popular in the Europe during the Renaissance (1350 to 1600 AD). Some of the most important advances in the history of plant propagation were made during the last 100 years. One such mile stone was discovery of auxin, which revolutionized the nursery industry. This was evident by release of two commercial products, *viz.,* Hormodin and Rootone for rooting of cuttings, within five years of the discovery of auxin. Now, very frequently a new dimension is being added to the field of plant propagation in general and horticultural plants in particular. For instance, invention of glass house; mist propagation; hybrid seed production technology; micro grafting; rooting co-factors in the rooting of cuttings; micropropagation; epicotyl grafting in mango; stooling in guava, mango and litchi; side veneer grafting in mango; use of juvenile parts for easy rooting, bottom heat technique in rooting of cuttings and emphasis on sanitation and disease control in nursery are few new entries in the field of propagation of horticultural crops in the last century.

Propagation can be defined as the science and art of multiplication of plants by either sexual or asexual means. It is the process of multiplying the numbers of a plant, perpetuating a species or maintaining the youthfulness of a plant. The first and foremost objective of a propagation method should be to produce individuals that are identical to mother or original plant. Thus, a successful propagation method is one, which transmits all the desirable characters of a mother plant to the offspring. If a method is unable to do it, that is considered as an undesirable method. Therefore, sexual reproduction of fruit trees is not an often used method in horticulture. Sexual propagation, primarily, finds its application in breeding programme so as to conserve the diversity of the existing gene pool and to develop new varieties. From production of quality planting materials point of view also, this method cannot cater the requirements for production quality as well as quantity due to various reasons like single seed per fruit, poor germination, and specific treatment requirement before germination etc. The new variant, as an outcome of the sexual reproduction, carries unpredictable characteristics of its own, which may include the tendency to revert to a wild-like state of the species. However, some of the known fruits such as chest nut keep intact their specific characteristics of the variety independently from the way, which was used for propagation. Irrespective of propagation methods, such fruits inherit the characteristics of the variety. Most of the present day horticultural plants are selected from the wild species, which are the resultant of natural hybridization and mutation and if these are not multiplied by the suitable methods, these may lost their identity and all efforts made by the breeder will go in vain. Thus, only desirable or recommended method of propagation should be tried.

Methods of propagation of fruit plants: An overview

Sexual propagation

Seeds: Propagation by seed is considered as the most important method of multiplication of most of the annuals and many perennial fruit plants. By and large, propagation of plants by seeds is considered as sexual propagation; though, not all embryos develop from sexual processes e.g. some embryos develop through apomixes. Fruit plants are propagated by seeds either (i) for raising commercial orchards in case of some species, or (ii) to develop new varieties, or (iii) for raising rootstocks for budding and grafting purpose etc.

Asexual propagation

Principally, plant can be asexually reproduced either by using part of two or more plants in a union (grafting) or parts of the same plant (rooting). All asexual propagation techniques belong one of the two categories.

Propagation by apomictic seedlings

For example, citrus, mango, apple etc.

Propagation by vegetative structures

a. Propagation by division

Division of tuberous roots: *Ampelocissus latifolia*, turmeric, dahlia, bigonia etc.

Division of rhizomes: Banana, blueberry, iris, oxalis etc.

Division of stem tubers: Potato, Jerusalem artichoke, caladium etc.

Division of stock: Gooseberry, currant, strawberry etc.

b. Propagation by separation

Offsets: Date palm, banana, pineapple.

Bulbs: Onion, daffodil, Bird of Paradise, tulip, tuberose, lily, narcissus, hyacinth etc.

Corms: Gladiolus, freesia, crocus etc.

c. Propagation by suckers

Pineapple, banana, raspberry, blackberry, *kair, pilu* (root sucker), chrysanthemum etc.

d. Propagation by runners

Strawberry

e. Propagation by stolon

Doob grass

Propagation by cuttings

a. Stem cuttings

Herbaceous cuttings: Defenbacia, dahlia, chrysanthemum, begonia, coleus, carnation etc.

Softwood cuttings: Apple, pear, peach, plum, apricot (under high humidity conditions), Juniper, hollyhock, azalea etc.

Semi-hardwood cuttings: Mango, guava, jackfruit, lemon, rhododendron etc.

Hardwood cuttings

Simple cuttings: Grape, fig, pomegranate, mulberry, kiwifruit, olive, quince etc.

Torn cuttings: Quince (dwarf)

Hammer cuttings: Gooseberry

Truncheons: Fig

b. Leaf bud cuttings

Blackberry, lemon etc.

c. Leaf cuttings

African violet, sansevieria, bigonia, bryophyllum etc.

d. Root cuttings

Blackberry, red raspberry, bread fruit, wood apple, fig, cherry, apple, pear etc.

Propagation by layering

a. **Tip layering:** Trailing blackberry, raspberry etc.
b. **Simple layering:** Grape, lemon, ornamental climbers, bougainvillea etc.
c. **Stooling/mound layering/ bank up layering:** Apple rootstocks, guava, mango, litchi, gooseberry, hazel nut etc.
d. **Serpentine layering:** American grape etc.
e. **Trench layering:** Apple rootstocks, cherry, plum etc.
f. **Air layering:** Lemon, guava, litchi, macadamia nut etc.

Propagation by grafting

a. Whip grafting: Many fruit plants

b. Cleft grafting: Avocado, apple, pear, plum, camelia etc.

c. Bark grafting: Different fruit plants

d. Bark grafting: Different fruit plants

e. Side grafting: Mango, avocado etc.

f. Inarching: Mango, sapota, guava, litchi etc.

g. Veneer grafting : Mango

h. Stone/epicotyl grafting: Mango

i. Top working: Mango, *ber*, cashewnut, mulberry etc.

j. Softwood grafting: Cashewnut, mango

k. Bridge grafting: For repairing of the damaged fruit plants.

Propagation by budding

a. T-budding: Citrus, plum, peach, cherry, *ber*, rose etc.

b. Patch budding: Walnut, pecan nut, Indian rubber tree etc.

c. Chip budding: Mango, grape etc.

d. Ring budding: Peach, plum, *ber*, mulberry etc.

Micropropagation

a. Cell culture: Citrus

b. Meristem culture: Banana, strawberry, guava etc.

c. Shoot tip culture: Citrus, papaya, grape, strawberry, orchids, asparagus, lily, gladiolus etc.

d. Embryo culture and embryo rescue: Grape, peach, plum, apricot, citrus, rhododendron etc.

e. Anther culture: *Datura*

* * *

2 Sexual Propagation

Sexual reproduction involves meiotic cell division, which ultimately produce progeny with new or differing genotypes relative to their male and female parents. It is also called as seed propagation. It entails the union of the pollen from the male sex organ of flower (androecium) with the egg from the female sex organ (gynoecium) in order to produce a seed through processes like pollination and fertilization. The seed is made up of three parts *viz.*, the outer seed coat, which protects the seed; the endosperm, which is a food reserve; and the embryo, which is the young plant itself. The seeds on germination give rise to individuals with a new genotype. The plants produced through seeds are called as seedlings. When used to raise commercial orchards, sexual propagation has certain advantages and disadvantages over vegetative propagation, which are discussed hereunder;

Advantages

1. Compared to vegetative propagation, propagation through seed is relatively a cheaper and simple method of obtaining a large number of plants. Even, in some fruit plants like papaya, this is the most popular method of propagation.
2. In general, seed raised plants have long life and better root system (tap root); thereby, provide better anchorage as compared to vegetatively propagated plants, which have adventitious roots.
3. Usually, seedling plants are resistant to insect-pests and diseases than the vegetatively propagated plants. In addition, they can tolerate vagaries of adverse climatic conditions in a better way than the vegetatively propagated plants.
4. Generally, rootstocks for budding and grafting are raised through seeds, except clonal rootstocks.
5. The hybrids, derived from controlled hybridization programme, are first raised through seeds. Likewise, Sexual propagation sometimes leads to the production of chance seedlings, which may be superior to the parent plants.
6. Nucellear seedlings can be utilized to raise uniform plants. The mangosteen (*Garcinia mangostana*) fruit is developed through parthenogenesis and seedlings

obtained from the seeds of such fruits are similar to its mother plant and thus propagated commercially through seed. Similarly, polyembryony is a rule in some fruit *viz.*, citrus (except pummelo, Tahiti lime and Turanj) and mango (varieties like Kurukkan, Chandrakaran, Carabao, Goa and Vellaicolumban etc.)

7. It is easy to store seeds for longer time than the vegetative parts and can be transported easily to distant places. Moreover, storage of seeds involves limited space and time than vegetative parts of a plant.
8. Handling of seeds is easy as compared to vegetative parts of the plant.
9. Usually, not much technical skill is required for raising plants through seeds
10. Seed propagation is useful in producing virus free plants as most viruses are not transmitted through seeds.
11. With small quantity of seed, a large population can be raised.

Disadvantages

Propagation through seeds has certain limitations over asexual methods, which are discussed as under;

1. In majority of instances, seedling plants are highly heterozygous in nature. Thus, the seedling plants are not uniform in growth, yield and fruit quality parameters than vegetatively propagated plants.
2. Seedling plants are usually tall, vigrous and spreading types and pose problems for efficient management. Thus, the management cost involved in operations like harvesting, pruning and plant protection etc. is comparatively higher than the vegetatively propagated plants.
3. In general, seedling plants have longer juvenile phase and take more time to come into bearing than vegetatively propagated plants.
4. Usually, seeds in most of fruit plants loose viability very soon after extraction and thus have very low germination rate.
5. It is not possible to derive the beneficial effects of rootstock on scion variety with seeds.
6. Some fruit plants like pineapple, banana does not produce viable seeds and propagation through seeds is not possible in them.
7. Seed borne viruses exist in a number of fruit plants, as psorosis in citrus and mosaic in peach, cherry and almond. Thus, multiplication of these plants by seed is not feasible.
8. It is not possible to perpetuate the exact character of any superior selection through seeds.

9. Seed often require various treatment such as cold chilling, warm stratification, mechanical or chemical seed coat scarification, or some combination of these or other treatments as pre-requisite for germination to take place. Lengthy dormancy requirements may result in a more costly product or a longer construction schedule. Sometimes, unknown dormancy requirements may prevent propagation by seed entirely.

Significance of seed development

Seeds are, principally, a means of reproduction and most seeds are the product of sexual reproduction. Production of seeds by flowering plants, both angiosperm and gymnosperm, is considered to be a preadaptation for the survival and rapid dominance of such plants on earth. Besides, the enhanced sporophyte (diploid), reduced gametophyte (haploid), and a wide range of morphological adaptations (roots, leaves, the cuticule and stomata, to name a few) allow seed plants to survive in a range of habitats and dominate the terrestrial flora. The seed habit itself, in addition to vegetative traits such as the production of wood by a secondary meristem (cambium), contributed decisively to the evolutionary success of the gymnosperms and angiosperms.

Structure of the seed

A seed can be hailed as a meristematic axis, often associated with a storage tissue (endosperm) and enclosed by an enveloping series of membranes and sometimes stony shells, which collectively make the seed coat. Further, it can be defined as a mature, integumented megasporangium. There are some terms, which are necessary to be understood for better comprehension of seed system (Table 1).

Table 1. Some common terms associated with seed system.

Ovule	Structure that consists of the integument(s) surrounding the nucellus (megasporangium); unfertilized, immature seed precursor
Nucellus	Megasporangium; surrounds the megagametophyte; can develop into perisperm after fertilization
Megagametophyte	Female gametophyte, contains the female haploid egg cells (gametes) and several thousand (gymnosperms) or typically three to eight (angiosperms) other cells; the mature angiosperm megagametophyte is called the embryo sac
Integuments	One (gymnosperms) or two (angiosperms) outer layers of the ovule, having an apical opening (micropyle); develop after fertilization into the seed coat (testa)
Micropyle	Apical opening of the integuments; allows the pollen tube to enter the nucellus to release sperm for fertilization

Testa	Seed coat, derived from the integuments of the ovule; dead maternal tissue.
Endosperm	Arises from the fusion of a second sperm nucleus with the central cell nucleus of the embryo sac during *double fertilization*; nutritional tissue during seed development and in the mature seed of most angiosperms, in between testa and embryo. The genetic constitution of endosperm is 3n.
Perisperm	Derived from the nucellus after fertilization; maternal nutritional tissue in the mature seed of some angiosperms.
Ovary	Usually lower portion of the angiosperm pistil (carpel or fused carpels) containing ovules; fruits are mature ovaries; ovary tissue develops into the pericarp.
Pericarp	Fruit coat of angiosperm fruits, develops from the mature ovary wall and other flower tissues; surrounds the seed(s).

Seeds of different plants vary greatly with respect to size, form, external colour and markings, internal structure, and amount and nature of stored food, but most seeds consist of the following parts;

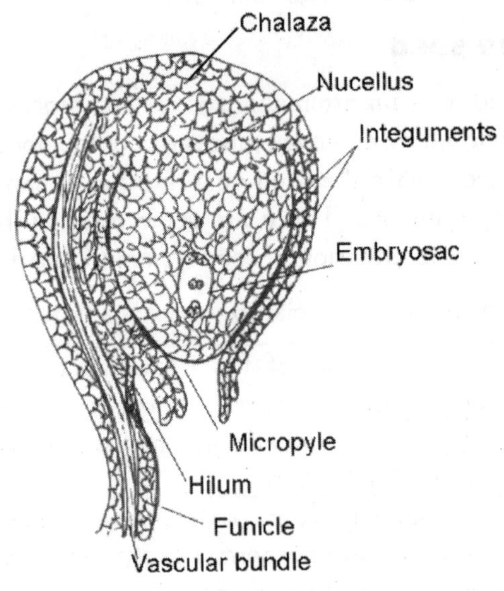

Fig. 1: Different features of a seed.

The seed coat is developed from the integument or integuments of the ovule. When two seed coats are present, the inner one is usually thin and delicate, and the other one is thick and tough, which is a very effective protective structure. The principal external features, which may be found on mature seeds, are the *hilum, raphae, micropyle,* and *chalaza* (**Figure 1**). The hilum is a scar, which marks the place where the seed broke from the stalk (funiculus). In those plants in which the ovules become curved, so that the micropyle lies close to the hilum, there may be a ridge composed chiefly of a bundle of vascular elements. This ridge, formed by the stalk of the ovule, and present only in ovules in which the funiculus is sharply bent at the base of the ovule, is called the *raphae*. *Chalaza* is described as the base of the ovule opposite the micropyle, where integument and nucellus are joined together. In seeds, which when mature are without endosperm, the endosperm has been used up by the developing embryo before the seed becomes fully developed. In seeds of this type, such as the common bean, pea, alfalfa, clover etc., the stored food is mostly in the cotyledons. Conspicuous examples of seeds with endosperm are castor oil seed, corn, wheat and other cereals. The rudimentary plant or embryo differs very greatly in appearance in different seeds, owing generally to differences in the form and relative development of the parts of the embryo. However, with few exceptions, all embryos are made up of the same organs. The organs of the embryo in most seeds are (Figure 2):

1. **The plumule**- a rudimentary shoot.
2. **The cotyledons**- seed leaves, sometimes only one.
3. **The hypocotyl**- part of the embryo between the attachment of the cotyledons and the upper end of the rudimentary root, usually very short in some seeds.
4. **The radicle**- rudimentary root.

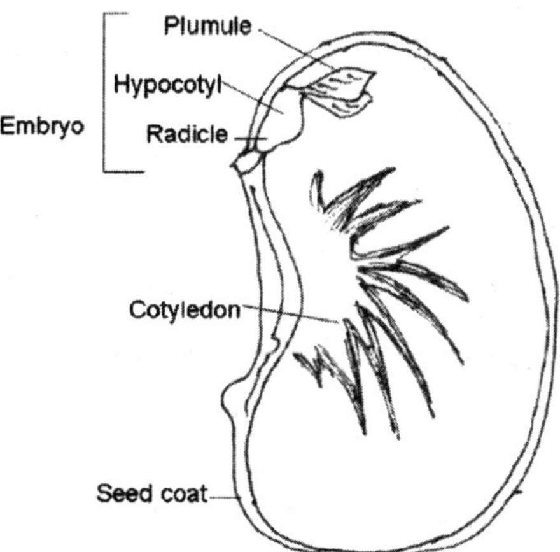

Fig. 2: Seed of common bean showing different organs.

In seed, the principal kinds of foods stored are chiefly carbohydrates (such as starch, hemi-celluloses and sugars), fats and proteins.

Seed dormancy

After extraction, seeds are either stored for some time or they are sown directly for raising new plants. However, many seeds fail to germinate due to cessation of growth after processing and placement in seemingly favorable growing conditions i.e. water, temperature and aeration are permissive for germination. In such seeds, depending upon their types, some morphological and biochemical changes are prerequisite before germination takes place. The suspension of growth in seeds could be attributed to external environmental factors (e.g. water, temperature, light etc.) and/or to internal limitations. The arrested seed growth due to external environmental factor/stimulus is referred as *quisence*, while the one mediated by internal barriers is termed as *rest phase*. Since, the ultimate outcome is arrested seed growth; therefore, both situations are put together to be termed in general as *dormancy*. In general, *seed dormancy* could be considered simply as a block to the completion of germination of an intact viable seed under favourable conditions. This block to germination has evolved differently across species through adaptation to the prevailing environment, so that germination occurs when conditions for establishing a new plant generation are likely to be suitable (**Figure 3**). Therefore, a diverse range of blocks (dormancy mechanisms) have evolved, in keeping with the diversity of climates and habitats in which they operate. Though, dormancy seems to be a challenge for plant propagators; however, it should be considered as a mechanism, evolved during evolution, to sustain the next generation (seeds) by plants. Nature has gifted these species by way of preventing germination prior to hostile environmental conditions that would otherwise kill young, tender seedlings after germination. Since, it is one of the major problems faced by the plant propagators involved in seed propagation; therefore, it is inevitable for the plant propagators and nurserymen to have fair knowledge of dormancy and measures to overcome it.

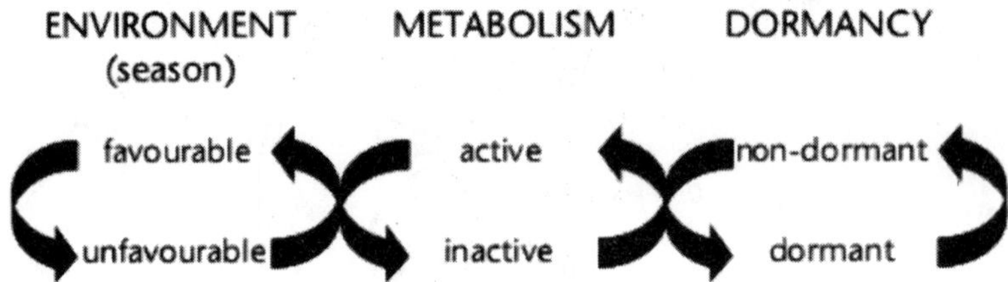

Fig. 3. Synchronization of plant metabolic activity and dormancy with the seasons. (Adapted from Hilhorst and Koornneef, 2007)

Dormancy: its kinds

In the literature, various kinds of dormancy have been reported. Depending on different species and mechanism, it is primarily of the following kinds;

1) Primary dormancy

In general, primary dormancy is imposed during seed and embryo development so that varying degrees and types of dormancy may be exhibited by the seed during the later stages of development. Primary dormacy has been grouped in to following classes;

A. Exogenous dormancy

Germination, which is marked by radicle protrusion, is the net result of opposing forces exerted by the thrust of the embryo against the mechanical restraint of the surrounding tissues. It is well established that the dormancy may be located in both the embryo and the enveloping tissues. The tissues surrounding the embryo can inhibit germination of seed by inhibiting water uptake, providing mechanical resistance to the expansion of embryo, limiting gaseous exchange, preventing leakage of substances from the embryo, supplying inhibitors to the embryo etc. Therefore, the exogenous dormancy is mainly attributed to the seed coat, which can exert physical, mechanical or chemical barriers. This could be of the following types;

a. Physical or seed coat dormancy: Many seeds possess a seed coat, which poses a mechanical restraint to embryonic growth. The seed coat or the seed coverings in such seeds may become hard, fibrous or mucilaginous during hydration and ripening, as a result, they become impermeable to water and gases, which prevents the physiological processes leading to the germination process. In addition, the layers of fleshy fruits may dry and become the part of seed covering. In drupe fruits (peach, plum, apricot etc.) these layers become hardened endocarp (pit or stone). In others, they become the shell (walnut, pecan nut etc.). The seeds having seed coat dormancy do not germinate until impermeable barrier is weakened or replaced. Though the embryo is quiescent (non-dormant) but covered by a water impermeable covering. The impermeability of a seed could also be due to a layer of thick walled palisade like *macrosclereid* cells, coated with a thin layer of waxy cuticular substances during seed maturation and ripening.

In nature, the impervious seed coat and its coverings are softened by the action of various agents of environment like, continuous weathering, mechanical abrasion, alternate freezing and thawing, attack by soil microorganisms (fungi and bacteria) or passage of the seeds through the digestive tracts of animals. Softening of seed covering by fungi and bacteria is more effective above 50°F. As a result of these actions, the seed coat softens and become permeable to water and gases, which facilitates easy germination of the seeds.

b. Mechanical dormancy: Sometimes, the seed enclosing structures, such as, shells of walnut, pits of stone fruits, and stones of olive and *ber* are so hard that they do not allow the embryos to expand during germination. In such cases, water may be absorbed but difficulty arises in the cementing material that holds the dehiscent layers together. In general, hardening of seed coverings is never a sole cause of dormancy in any species but the hardening in combination with other factors, may delay seed germination. For example, the seed coat (pits of stone fruits and shells of nuts) are apparently permeable to water and gases, but have very high mechanical resistance to allow expansion of embryo. Thus, the germination in such seeds does not occur until and unless the seed coats are softened.

Such mechanical resistance of seed covering is decreased by creating moist and warm conditions during storage or by microbial activity.

c. **Chemical dormancy:** Certain inhibitory chemicals e.g. phenols, coumarin and abscsic acid may get accumulated in fruits and seed covering tissue during development and remain with seed after harvesting and later act as germination inhibitors. This phenomenon is quite common in fruits like citrus, stone fruits, apple, pears, grape etc. Though, it is very difficult to establish a direct relationship between inhibitors and seed germination but there are instances (e.g. Iris), where germination of seeds has been improved or stimulated by leaching of inhibitors with water, or removing seed coverings or both.

B. Endogenous dormancy

Morphological dormancy: Morphological dormancy is evident in seeds with embryos, which are underdeveloped (in terms of size), but differentiated (e.g. into cotyledons and hypocotyl-radical). These embryos are not (physiologically) dormant, but simply need time to grow and germinate. Some plants of Temperate Zone plants like holly and snowberry also have rudimentary embryos. Likewise, seeds of Chinese gooseberry and *Annona* require warmth of 2 and 3 months, respectively for adequate germination. Sometimes germination-inhibiting chemicals (inhibitors) also become active at high temperature, making the problem much more complex. The germination in seeds having such problems can be improved by i). Exposure of the seeds to a temperature of 15°C or below, ii). Exposure of seeds to alternate temperatures, and iii). Treatment of the seeds with chemical additives like potassium nitrate or gibberelic acid.

b. **Physiological dormancy:** Seeds of some species exhibit physiological dormancy. It has been classified in to non-deep dormancy, intermediate dormancy and deep dormancy as discussed below:

i) **Non-deep physiological dormancy:** Embryos excised from these seeds produce normal seedlings; GA treatment can break this dormancy and, depending on species, dormancy can also be broken by scarification, after-ripening in dry storage, and cold or warm stratification. This type of dormancy may be due to heat sensitivity (thermo-dormancy) in which temperature has direct influence on seed germination or due to light sensitivity (photo-dormancy) in which seeds of some species require light or darkness to germinate.

Photo-dormancy: Seeds, which require a certain period of either light or darkness for germination, are called a *photo-dormant.* The basic mechanism of light sensitivity in seeds involves a photochemically reactive pigment, called as *phytochrome,* widely present in plants. When imbibed seeds are exposed to red light (660 nm), the phytochrome changes to red form (P_{fr}), thereby stimulating the germination process. However, when seeds are exposed to far-red light (730 nm), P_{fr} is changed to P_r, which inhibits germination process. Similarly, the germination in seeds is inhibited or favoured by the length of the day. For example, seeds of *Chenopodium album* remain dormant if its plant are exposed to long days but have no dormancy if raised under short day conditions.

Thermo-dormancy: Some seeds have specific temperature requirements for their germination, otherwise, they remain dormant. Such seeds are called as *thermo-dormant.* For example, seeds of lettuce, celery and pansy etc. don't germinate if the temperature is below 25°C.

The control of physiological dormancy lies in physiological activities and living seed coverings. Removal of seed coat and abrasion may bring about germination in many species. This mechanism is purely due to semi-permeability of seed coverings, which allows water to enter on removal of coverings but not the gases.

ii) Intermediate and deep physiological dormancy: Seeds with intermediate and deep physiological dormancy are commonly found in trees, shrubs and some herbaceous plants of temperate region, where seeds over winters in soil and germinate in spring under natural conditions. Such seeds are characterized by a requirement of 1 to 3 months of chilling in imbibed and aerated state, commonly called as *moist chilling.* This requirement of the seeds of temperate zone plants has led to the development of world known horticultural practice of *stratification* to meet the moist chilling requirement of seeds, which is required to overcome the physiological dormancy. In stratification, the seeds are placed between layers of moist sand or soil in boxes or in ground and are exposed to chilling temperature either out doors or in the refrigerators. For successful stratification, the seeds should be stored in a moist, aerated medium at chilling temperature for certain time.

To overcome embryo dormancy, the following pre-requisite conditions are necessary:

1. Imbibition of moisture by seed
2. Exposure to chilling temperature
3. Proper aeration and
4. Certain time period

Moisture : It is very important that the seeds have imbibed sufficient moisture in order to assure proper after ripening of the dormant embryos. In general, the process of imbibition is fast in absence of any barrier like bony or hard endocarp. Once imbibition completes, the seeds swell and seed coverings split open. For proper stratification, dry dormant seeds must absorb moisture by imbibition to around 50 percent.

Temperature : It is the single most crucial factor influencing the process of after ripening.

Usually, temperature above freezing i.e. 2-7 °C is the most effectual for moist chilling process. Temperature variation, both below or above, may severely impact the after ripening rate. At a higher or lower temperature, seeds usually fail to germinate and this may lead to induction of secondary dormancy in seeds.

Aeration : Though, the amount of O_2 needed during stratification is related to temperature, but proper supply of O_2 during seed stratification is absolutely necessary for proper after ripening process of the seeds. The amount of O_2 required for after ripening process depends on temperature, moisture content of imbibed seeds and type of the seed. In general, the O_2 requirement at chilling temperature is low but if the temperature is more, the O_2 demand also increases.

Time period : The time required for after ripening process is dependent on species, condition of seed, the management of seed bed, seed handling and chilling requirement of the seeds. In general, seeds of most woody perennials require 1 to 3 months time for after ripening, though certain species require even 5 to 6 months also.

Seeds having intermediate physiological dormancy usually require chilling stratification to overcome dormancy, whereas seeds having deep physiological dormancy require a relatively longer period (7-8 weeks or even more) of moist chilling stratification to overcome dormancy. In general, seeds having deep physiological dormancy give rise to normanl seedlings. However, typical un-chilled seeds may give rise to "physiological dwarfs" (seedlings grow out of dwarfism after sometime).

c) Epicotyl dormancy: Seeds of some species have different stratification requirement for their radicles, hypocotyl and epicotyl and this type of dormancy is called as *epicotyl dormancy*. On the basis of chilling requirement for epicotyl, different plant species are divided into two sub-groups. In one group, seeds germinate initially during warm period of 1-3 months and produce roots and hypocotyl but these require 1-3 months chilling for epicotyl to germinate as in lily (*Lilium*), *Hepatica antiloba* and peony (*Paeonia*) etc. In second group, seeds require a certain amount of chilling period to after ripen the embryo, followed by a warm period for the roots to grow and again a chilling period to stimulate shoot growth as in *Trillium* and certain other perennials of temperate climate.

C. Double dormancy

It is evident in seeds with water-impermeable coats (as in physical dormancy) combined with physiological embryo dormancy. Therefore, it is called as 'double dormancy' or combinational dormancy. It may be exo-endo dormancy that combines seed coat dormancy and intermediate physiological dormancy (as in some tree legumes) or morphophysiological dormancy in which there is development of rudimentary embryo along with physiological dormancy. For proper germination, both the blocking conditions must be modified to allow water to penetrate the embryo and after ripening of the embryos, following chilling stratification. This type of dormancy is found in trees and shrubs producing seeds with hard seed coats, grown primarily in cold winter area. To overcome the problem of double dormancy, the seeds are first treated in a way to soften the seed coat and then seeds are chilled as they overwinter.

2. Secondary dormancy

Dormancy can also be induced in mature, already dispersed, nondormant seeds (known as *induced* or *secondary dormancy*) by environmental conditions, which are unfavorable for germination, for example, anoxia, unsuitable temperature or illumination. This type of the dormancy develops after the imbibition of water by the seed but before the emergence of the radicle, primarily, due to exposure of non-dormant seeds to unfavourable environmental conditions like high and too low temperature, prolonged darkness (skoto dormancy), prolonged white light (photo dormancy), water stress and deficiency of oxygen and carbon-dioxide. Seeds can be released from secondary dormancy by chilling or light treatments and in some cases by treating with gibberellic acid according to the dormancy conditions.

Different factors involved in the development of primary and secondary dormancy in seeds are presented in **Figure 4**.

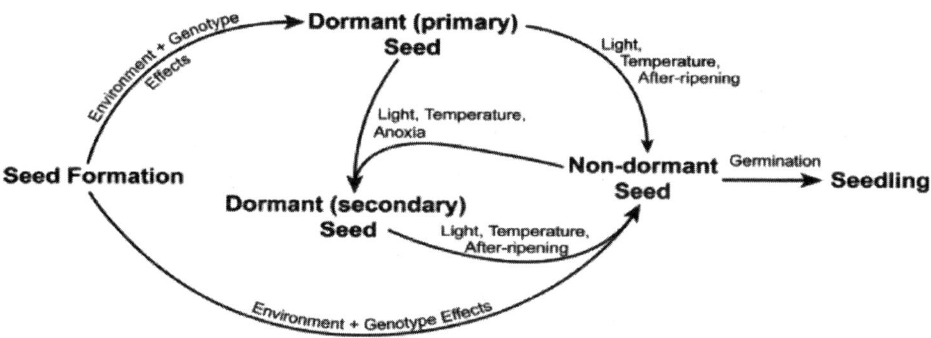

Fig. 4. The development of primary and secondary dormancy in seeds and factors influencing its inception, degree and termination. (Adapted from Bewley and Black, 1994).

Seed dormancy and its relevance to plants

Seed dormancy has an enormous relevance to the existence of many plant species from ecological perspective. It is an evolutionary adaptation for persistence and survival by some plant species in which germination is delayed even under seemingly favourable conditions of aeration, moisture and temperature. Thus, nature has bestowed these plants because it is advantageous to them in the following ways;

1. Seed dormancy controls germination timing in response to the seasons and plays an important role in seed plant evolution and adaptation to climatic changes. Due to dormancy, seeds germinate only when environmental conditions become favourable for seedling survival. In plants growing in arid regions, seeds can remain viable but dormant until sufficient water is available. The availability of sufficient water through rainfall ensures better survival of seedlings. Similarly, the impermeability of seed coat to water in some arid species also helps them in persisting in such hostile environments. In order for these seeds to imbibe water and germinate, the seed coat must be broken. However, permeability to water gradually occurs over a long period of time. This results in germination of only few seeds at a given time. Therefore, it is practically impossible for the entire species to be obliterated during the vulnerable seedling stage due to some adverse environmental conditions.
2. Seed dormancy facilitates dispersal of seeds effectively. It is particularly helpful in seeds having wings or special structural adaptations, which are only modified if they pass through the digestive tract of animals or birds.
3. Seed dormancy can also synchronize germination at a particular time of the year. For example, spring germinated seedlings of plants of temperate zone plants have sufficient time to grow and develop and that of summer germinated seedlings to acquire sufficient growth to face coming winter.

Relationship between hormones and seed dormancy

The development of a seed is characterized by an extensive change in levels of different growth promoting and inhibiting hormones (**Figure 5**). The commonly associated growth

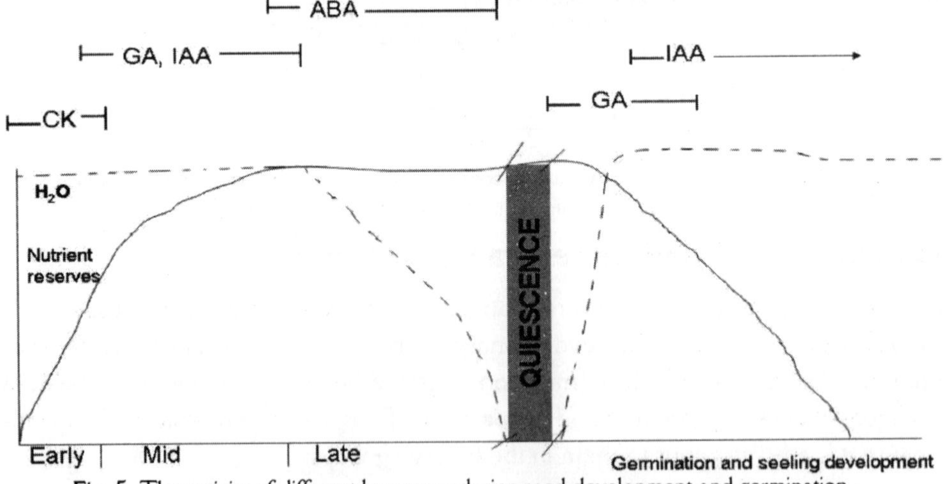

Fig. 5. The activity of different hormones during seed development and germination.

promoting hormones are gibberellins, cytokinins and ethylene, whereas abscisic acid acts as an inhibitor. Hormonal application has been proven to be most successful aid in germination process of many seeds. The success; however, depends upon species, variety and the stage of dormancy or germination.

In most seeds, cytokinins levels are highest during the very initial stages of embryo development when the rate of cell division is also highest. With the level of cytokinins declining and increase in the levels of gibberellins and IAA, the seed enters a period of rapid cell development. A negligible or absence of ABA is noticed during the early stages of embryogenesis. However, during the latter stages of embryo development as the levels of gibberellins and IAA begin to decline, ABA levels begin to rise. Usually, ABA levels attain peak during the maturation stage when seed volume and dry weight also reach a maximum. Maturation of the embryo is characterized by cessation of embryo growth, accumulation of nutrient reserves, and the development of tolerance to dessication. Since, ABA level increases with the maturation of fruit and; therefore, it is involved in the prevention of vivipary (precocious germination of developing embryos in the ovule) and induction of dormancy in many plant species. Its level is quite high in seed coats of dormant peach, walnut, apple and plum, which usually disappears as and when the seeds are stratified. Therefore, a crucial role for ABA has been identified in inducing seed dormancy. Factors determining spatial and temporal ABA content and sensitivity patterns in seeds promote induction of dormancy and probably its maintenance, and discourage dormancy release and germination. Gibberellins releases dormancy, promote germination and counteract inhibitory effects of ABA directly or indirectly. Gibberellins are required for embryo cell elongation, for overcoming seed coat restrictions to germination of non-dormant and dormant seeds, and for inducing endosperm weakening. Brassinosteroids and ethylene also counteract the inhibitory effects of ABA on seed germination, but in most species, they appear to act after dormancy has been released by gibberellins. The effect of interaction of different hormones and other factors on germination of tobacco seed (as a model crop) has been illustrated in **Figure 6**.

Phases of seed dormancy

The period of dormancy in seeds follows four different phases *viz.* induction phase, maintenance phase, trigger phase and germination phase, which are briefly described hereunder;

1. **Induction phase:** The dormancy in seeds is chiefly induced by several environmental factors such as light, temperature, humidity etc. For instance, hot and dry weather during seed maturation helps in the development of hard seed coat in clover seeds. However, the seeds have soft coats if they mature and ripen during rainy season. In addition, the balance between the growth promoters and the inhibitors has also some control on induction of seed dormancy. Seed dormancy in

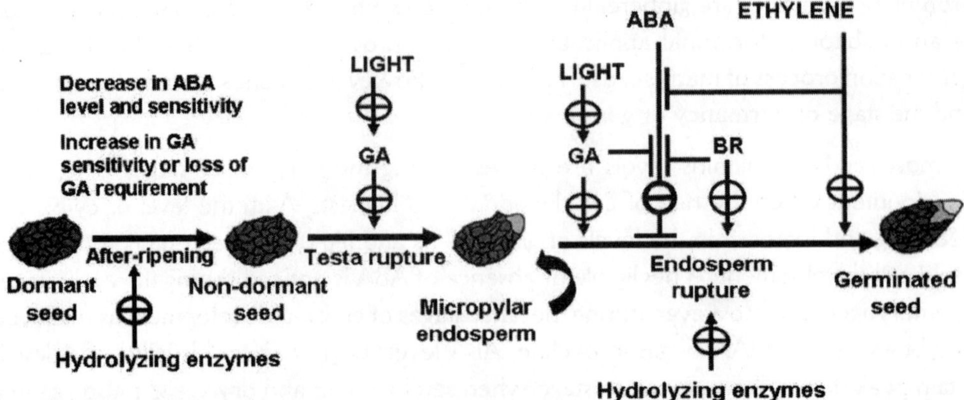

Fig. 6. Hormonal interactions during tobacco seed dormancy release and germination and their effects on testa and endosperm rupture. Dormancy release involves the light/gibberellin (GA) pathway and results in testa rupture. The light/GA pathway also promotes endosperm rupture. Hydrolyzing enzymes are also induced by the light/gibberellin (GA) pathway in the micropylar endosperm and facilitate endosperm rupture and radicle protrusion. Endosperm-speci?c hydrolyzing enzymes and endosperm rupture are inhibited by abscisic acid (ABA). The light/GA pathway also counteracts ABA effects by promoting ABA degradation. Ethylene and brassinosteroids (BR) counteract ABA effects and promote endosperm rupture, but do not affect testa rupture. BR and light/GA promote tobacco endosperm rupture by distinct signal transduction pathways. In addition, seed after-ripening is associated with a decrease in ABA content and sensitivity, and an increase in GA sensitivity or loss of GA requirement. A 'plus' sign means promotion and a 'minus' sign inhibition (Adapted from Kucera *et al.*, 2005; Finch-Savage and Gerhard, 2006).

many species is induced if the level of inhibitors (ABA) is more than the level of promoters (auxins, gibberellins and cytokinins).

2. **Maintenance phase:** In this phase, the promoter-inhibitors balance has shift towards inhibitors, which cause certain metabolic blocks. This phase can be shortened or lengthened artificially through exogenous application of growth promoters or inhibitors. Usually, pre-treatment of seeds either with a promoter or inhibitors at 40-50°C for 2 to 3 hours may shorten the rest period to some extent.

3. **Trigger phase:** In trigger phase, the promoter-inhibitors balance shift towards promoters by the environmental triggers. Stratification, scarification, exposure of seeds to light and leaching of inhibitors are some treatments, which may trigger the dormancy in many seeds.

4. **Germination phase:** A non-dormant embryo is characterized by increased growth potential, the capacity for cell extension growth and the ability to induce the release of dormancy. During this phase, the seeds germinate under favourable environmental conditions. Water is absorbed by seeds, they get imbibed, as result of which, the protoplasm becomes active. The metabolic block created by one or other factors is removed and the metabolic activity of the seeds increases due to

the increased enzymatic and hormonal activities, which initiate the germination process.

Treatments to overcome seed dormancy

Various treatments can be employed by the propagators to overcome seed dormancy. A brief discussion on various treatments has been given below:

1. **Softening seed coat and other seed coverings:** Softening of seed coat and other coverings of seed helps in better absorption of water and gases, which results in better germination of the seeds. It can be done by a process called a *scarification* as discussed below:

   a) **Scarification:** It is the process of scratching, breaking, chipping, altering, nicking or softening the hard seed coat and other coverings of the seeds to make them permeable to water and gaseous exchange. Scarification can be done either by mechanical means, chemically and by hot water treatment as follows:

      i) **Mechanical scarification:** This is a very simple and effective mean of scarification. Seeds are scratched or injured by rubbing against rough surface or sand paper. For large scale scarification, mechanical scarifies are used. The seeds should be dried after mechanical scarification and should be properly stored or planted immediately, as these become liable to attack by pathogens.

      ii) **Acid scarification:** Dry seeds are treated with concentrated sulphuric acid in the ratio of 1:2 in glass or wooden containers. Seeds should be continuously stirred in order to avoid injury. The treatment time depends upon the kind of seeds, varying from 15 minutes to 3 hours. After the treatment, the acid should be poured off and seeds are washed with running water for 10-15 minutes. The treated seeds may be sown immediately when wet or can be dried and stored for later sowing.

      iii) **Hot water scarification:** Hot water treatment is also a useful measure to overcome seed dormancy in some plant species. In this treatment, the dry seeds are placed in to four to five times their volume of water, maintained at 77 to 100°C. The source of heat is immediately removed and the seeds soaked in gradually cooling water for 12-24 hours. The unswollen seeds are removed and discarded, whereas, the imbibed seeds can be sown immediately.

      iv) **Warm moist scarification:** Seeds are placed in moist-warm medium for many months to soften their seed coats through microbial activity. This treatment is highly beneficial in seeds having double dormancy.

v) **High temperature scarification:** Seed coats of certain species are modified only if they are subjected to high temperature treatment. For example, the hard seeds of some gymnosperms (e.g. *Pinus radiata*) germinate extensively after a forest fire. Perhaps, the high temperature of fire may be responsible for the melting of resin that usually seals the cone. The resin of the cone melts due to high temperature of fire, which breaks the seed dormancy.

vi) **Harvesting immature fruits:** In certain plant species, seeds are extracted pre-maturely to avoid the development of hard seed coats. Such seeds should be sown immediately after extraction.

vii) **Use of Charate:** The char from burned plant stems has been shown to be a good neutralizer of germination inhibitors in the seeds of several herbaceous species associated with chaparral fires. Golden yarrow (*Eriophyllum*) and *Phacelia* species have enhanced germination with the addition of a small amount of chamise (*Adenostoma fasciculatum*) charate to the sown seeds.

2. **Stratification:** It is a world known horticultural practice in which imbibed seeds are subjected to a certain amount of chilling temperature (5-10°C) to after ripen the embryo. Usually, the term moist chilling is used as a synonym to stratification. However, in some species of temperate region (e.g. fringe tree), which have epicotyl dormancy, require warm moist chilling for several months, followed by moist chilling to overcome dormancy. In general, stratification can be done by different means as follows:

i) **Refrigerated stratification:** The seeds are first soaked in warm water for 12-24 hours and when the seeds are fully imbibed, these are mixed in the ratio of 1:3 with a sterile medium, which retains moisture and provides aeration (peat, vermiculite, sand, sphagnum moss etc.). These seeds, mixed with medium are kept either in boxes, cans or jars with perforated lids and are then subjected to a temperature of 0-10°C for 4-6 months depending upon species. The boxes must be periodically examined and the medium may be remoistened immediately if the seeds are dry. This treatment is helpful in overcoming seed dormancy in most of the temperate fruits like apple, pear, cherry, plum etc.

ii) **Controlled outdoor stratification:** If refrigerated storage facilities are not available, outdoor stratification may be employed either by storing the seeds outside in deep pits or in raised seed beds, enclosed in wooden frames. The natural rainfall and winter chilling are thus used to meet the temperature and moisture requirements. In outdoor stratification, it is likely that the seeds are destroyed by excessive rains, freezing, drying or rodents. Hence, proper protective measures should be taken.

iii) **Outdoor planting:** Seeds of some species require cold treatment for after ripening. Such seeds may be sown directly in the seed beds so that these get proper moisture and other natural conditions during winter period. The seeds, in general, germinate quickly during spring, when the soil temperature begins to rise. It is a common observation that seeds of stone fruits (plum, peach and cherries), germinate promptly, if sown in early summer, mainly because these get enough period of warmth before the onset of chilling.

3. **Leaching of inhibitors:** It is a known fact that certain inhibitors and phenolic compounds are present in seed coverings of many species, which inhibit germination. Thus, soaking seeds in the running water for 12-24 hours or placing them in water for few hours helps in leaching off the inhibitors and phenolic compounds, which facilitates easy germination of seeds.

4. **Pre-chilling:** In seeds of certain plant species, dormancy can be overcome by pre-chilling treatment. In this treatment, the soaked or imbibed seeds are kept at a temperature of 5-10°C for 5-7 days before sowing. After this, the seeds can be sown in the field immediately.

5. **Pre-drying:** Pre-drying is also an useful practice in some seeds to overcome seed dormancy. In this treatment, the dry seeds are subjected to a temperature of 37-40°C for 5-7 days before sowing. After the treatment, the seeds can be sown in the field.

6. **Hormonal treatments:** Various hormones are known to play a vital role to overcome dormancy of the seeds. However, GA_3 is commercially used for breaking seed dormancy in various kinds of seeds. The concentration of GA_3 depends upon kind of seed but usually; a concentration of 200-500 ppm is most widely used. Cytokinin is another group of hormones used for breaking physiological seed dormancy and for stimulating germination in seeds of many species. Kinetin and BA (6-benzyl aminopurine) are commercial preparations of cytokinins used for overcoming seed dormancy. Soaking seeds in 100 ppm solution of kinetin for 3-5 minutes in highly effective concentration for overcoming dormancy in many seeds. The treatment time and treatment, however, vary from seed-to-seed. Ethylene also stimulates germination in seeds of some species.

7. **Treatment with chemicals:** Some compounds other than hormones are also known to break dormancy but their role is not clear. Thiourea is one example known to stimulate germination in some kinds of dormant seeds. For this, soaked seeds are put in 0.5 to 3 per cent solution of thiourea for 3-5 minutes. Afterwards,

the seeds are rinsed with water and are sown in the field. Similarly, potassium nitrate and sodium hypochlorite also stimulate seed germination in many plant species.

8. **Seed priming:** Seed priming refers to the procedures followed to overcome dormancy in freshly harvested seeds. Most widely used seed priming procedures are osmo-conditioning, infusion and fluid drilling.

    i. **Osmo-conditioning:** In this treatment, the seeds are placed in a shallow layer in a container having 20-30 per cent solution of polyglycol (PEG). The seeds are then incubated at 15-20°C for 7-21 days, depending upon seed size and plant species. Different hormones and fungicides can be added in PEG solution to protect the seeds from pathogens. After it, the seeds are washed and dried at 25°C and are stored until use.

    ii. **Infusion** : the hormones, fungicides or insecticides and antidotes are infused into dormant seeds through organic solutions. The seeds are placed in acetone or dichloromethane solution containing chemical to be infused for one to four hours. Afterwards, the solvent is allowed to evaporate and seeds are dried slowly in a vacuum desiccator for 1-2 hours. These seeds absorb the infused chemical directly in to the embryo when soaked in water. The treated seeds may be directly sown in the field/nursery.

    iii. **Fluid drilling** : In fluid drilling, the seeds are suspended in a special type of gel before sowing. Now-a-days different types of gels are available in the market but sodium alginate, guar gum, and synthetic clay are most widely used in fluid drilling.

Process of seed germination

A seed is a ripened ovule, which consists of an embryo and stored food supply (embryo sac), surrounded by protected seed coverings. The activation of metabolic machinery of seed embryo is the first and foremost step to initiate the seed germination process. Thus, seed germination is the process of reactivation of the metabolic activity of the seed embryo, resulting in the emergence of radicle (root) and plumule (shoot), thus leading to the production of a seedling or a young plant. The process of seed germination usually takes place after a period of dormancy and in presence of specific factors like optimun temperature, water and oxygen. In the early stages of growth, the embryo draws nutrients from the stored food material in the cotyledons or the endosperm of the seed. Later, new shoot/ leaves are developed, which produce their own photosynthetic system.

Stages of seed germination

The process of seed germination involves several consecutive but overlapping events like i) absorption of water, ii) initiation of cell enlargement and division, iii) increased enzymatic activity, iv) food translocation to growing embryo, v) increase in respiration and assimilation, vi) increase in cell division and enlargement, and vii) differentiation of cells in to tissue and organs of a seedling. The sequence of these events is not specific and one event may overlap the other. However, the entire process of germination can be divided in to following different stages:

1) Activation or awakening stage

Uptake of water by a seed is triphasic, with a rapid initial uptake (phase I, i.e. imbibition) followed by phase II, which is marked by a plateau. At phase III, a further increase in water uptake occurs only when germination is completed, as the embryo axis elongates and breaks through its covering structures. Cell elongation is necessary, and is generally accepted to be sufficient, for the completion of radicle protrusion; but cell division is not essential.

2. Translocation stage

Food materials like fats, carbohydrates or proteins are stored in the endosperm or in the cotyledons converted in to simpler forms and are translocated to the growing points of the embryo. All these conversions are regulated by metabolic activity of specific enzymes in a proper sequence. For example, fat and oils are first converted enzymatically to first to fatty acids and then to sugars. Storage proteins are first converted to amino acids and then to nitrogen, which are essential to growing seedlings. Starch present in many seeds as an energy source, is also converted to simple sugars.

3. Seedling growth stage

At this stage, the development of the seedling plant takes place from continued cell division in different growing points of the embryo, which is subsequently followed by the expansion of the seedling structures. This marks the final step in the germination of the seed, where the cotyledons are expanded, which are the true leaves of the seedling plant.

Types of seed germination

The radicle, the growing point of root emerges from the base of the embryo axis and the plumule, the growing point of shoot is at the upper end of embryo axis, above the cotyledons. The section of seedling stem above the cotyledons is called as *epicotyl* and below the cotyledons, is called as *hypocotyl.* Two types of germination are commonly found in cultivated plants. When the epicotyl grows first only the plumule is pushed out of soil. This is termed as *hypogeous* or *hypogeal* germination. This type of germination is very common in mango, custard apple etc. In the other type, the hypocotyl grows first and pushes the cotyledons and other parts of the seed out of soil. It is termed as *epigeous* or *epigeal* germination. This type of germination is common in beans, gourds, castor, tamarind and onion etc.

Factors affecting seed germination

Many factors affect seed germination, which can be broadly grouped as external factors, internal factors and storage factors, which are described briefly as under:

A. External factors

1. **Water:** The role of water in germination is of utmost importance. The process of seed germination starts only after water absorption. The uptake of water by seeds is called *imbibition*, which leads to the swelling of seed followed by the breaking of the seed coat. When the seed imbibes water, hydrolytic enzymes are activated, which break down the stored food resources in seeds (such as starch, proteins, or oils) into simple compounds for further use by growing embryo. Generally, the seed germination does not occur at moisture below 40 to 60 per cent. The rate of water uptake is influenced by many factors like seed size, nature of the seed coat, prevailing temperature, and available moisture contents etc. But in all, it is essential that the optimum moisture should be available in adequate amount to initiate the process of germination.
2. **Temperature:** Temperature affects seed germination percentage and rate. Usually, every plant species has certain optimum temperature for germination, below and above of which, the germination is inhibited. Thus, the optimum temperature is the one at which the largest number of seeds germinate at the highest rate. The optimum temperature range for seed germination in most plant species is between 25 to 30^{0}C, although seeds of certain vegetables (cabbage and carrot) can germinate at very low (3-4^{0}C) and at very high temperature (40^{0}C) and seeds of some species fail to germinate at a temperature higher than 25^{0}C. Temperature also regulates the subsequent seedling growth.
3. **Aeration:** Germinating seeds respire at a very high rate. O_2 is required by the germinating seed for metabolism. Oxygen is used in aerobic respiration, the main source of the seedling's energy until it grows leaves. Some seeds have impermeable seed coats that prevent oxygen from entering the seed, causing a type of physical dormancy, which is broken when the seed coat is worn away enough to allow gas exchange and water uptake from the environment. Germination rate is always poor in over watered or poorly drained seed beds.
4. **Light:** Seeds of certain plant (e.g. *Allium, Amaranthus*) have specific requirement for light and will not germinate in its absence. Such seeds are planted very shallow in the seed beds. The light requirement depends upon the age of seed, degree of imbibition, temperature, day length, level of hormones etc. Red light is considered to stimulate germination in seeds. In contrast, seed germination in *Nigella, Allium, Phlox, Amaranthus* and pansy is retarded by light. The seeds in which germination

is promoted by light are called *positively photoblastic* and the one in which germination is inhibited by light are called *negatively photoblastic.* Light requirement can be partially reduced by treating seeds with GA_3, thiourea or KNO_3. However, in general, the red light is considered as stimulative and far-red as an inhibitory agent for seed germination. It is generally accepted that a balanced light spectrum should be used, including both red and blue light to produce vigrous seedlings because red light alone produces tall, leggy plants. Under controlled conditions, white fluorescent lamps are useful because they provide adequate, well balanced light for plant growth.

5. **Interaction between temperature and light:** Both temperature and light combination have great influence on seed germination. It is particularly true with flowering annuals like alyssum, aster, ageratum, candytuft, amaranthus, balsam, gaillardia and celosia etc.
6. **Soil salinity or alkalinity:** Salinity prevents or retards seed germination in some sensitive species. Frequent light irrigations result in salt deposits in the upper soil layer particularly when the water is saline. Salinity may even damage the germinating seed. Higher concentration of the salts may even damage the growing seed. This problem can be avoided by using soils and water with low salt content, withholding or using lesser amount of fertilizers and by irrigating less frequently, if water is salty.
7. **Pathogenic organisms:** Presence of some pathogenic organisms on seed surface, in the growing medium or in irrigations water also inhibits germination process and subsequent seedling growth. 'Damping off' of the seedling in vegetables completely destroys the young seedlings. Proper care must be taken to treat the seeds with appropriate fungicides/insecticides before sowing.
8. **Mechanical injury:** It is likely that the seed or the germinating seedlings are injured by some tool during intercultural operations. Such seeds usually fail to germinate, resulting in poor germination of the seeds.

B. Internal factors

1. **Seed coat barriers:** Seed coat of some species is impermeable to water due to presence of palisade like macro-sclereid cells and the embryo is unable to develop and emerge due to moisture stress. Sometimes, seed coats are hard (walnut and some stone fruits) and the tender embryo can not rupture seed coat, which ultimately affects germination. Sometimes, hard coat prevents the proper exchange of gases, which is a pre-requisite for seed germination.
2. **Seed dormancy:** The failure of apparently ripe seeds to germinate in a suitable environment may be due to seed dormancy. Dormancy in seed may be due to many reasons, which has been described in the preceding pages of this chapter.

3. **Rest Period:** Post extraction rest period is sometimes responsible for poor germination as in case of fig.

C. Storage factors

Storage factors, which affect seed germination, are low moisture content of seed, low storage temperature, and modification of the storage conditions. Of these, the moisture temperature relationship has the most practical significance. Harrington (1972) has introduced a rule of thumb that seed lose half of its storage life for every 1 percent increase in seed moisture between 5 and 14 percent. Similarly, seed life is reduced to half for every 5 percent increase in storage temperature between 0 and 50° C. Thus, storage conditions of the seed had great bearing on the seed germination.

Seed viability

Seed viability refers to the ability of seed to produce an acceptable seedling under favourable conditions. Simple emergence of radicle does not mean that the seed is viable. When non-viable, the seed may be dead or diseased or may produce abnormal seedlings. Seed viability varies greatly in different kinds of seeds. For example, the seeds of mango, citrus, jackfruit and walnut are short-lived and remain viable only for a few days, whereas, seeds are long-lived in leguminous crops. Seed viability also depends upon the storage conditions. Some seeds are long-lived if stored at low temperature (0-4°C) and low humidity, whereas seeds of some species (litchi, mango, citrus, walnut, etc.) get desiccated and their embryos die, if stored under dry conditions. Seed viability can be expressed by germination percentage, which indicates, the number of seedlings produced by a given number of seeds. On the basis of storability of seeds in relation to moisture, seeds are classified into the following two groups:

a) **Orthodox seeds:** The seeds, which can be dried to a low moisture level and loose viability with the increase in moisture, are called as orthodox seeds. The seeds of *ber*, custard apple, date palm, fig, grape, guava, lemon, lime, mandarin, mulberry, papaya, passion fruit, peach, pineapple, plum, *phalsa*, and pomegranate are orthodox seeds.

b) **Recalcitrant seeds:** The seeds, which remain viable at relatively highly moisture content, and if dried below certain moisture level, these loose viability are called as recalcitrant seeds. The seeds of avocado, Barbados cherry, carambola, breadfruit, durian, jackfruit, litchi, mango, mangosteen, rambutan, walnut, and citrus are recalcitrant seeds.

Cryopreservation

It is the method of storage of material in liquid nitrogen maintained at a temperature of -196°C. Successful cryopreservation of material involves the application of chemicals known as cryopectants. The most commonly used chemicals are glycerol and DMSO (Dimethyl sulphoxide). This method has recently been used in preservation of embryos of coconut and jackfruit under *in vitro* systems.

Factors affecting seed viability

There are many factors, which influence seed viability but genetical factors, environmental conditions under which the seed is produced, mechanical injury to the seed during harvesting and the storage conditions of the seeds are usually reported to affect seed viability greatly. These factors have been described briefly hereunder:

a. **Mechanical injuries:** Mechanical injuries may damage certain seed structures, resulting in loss of seed viability. Such injuries usually occur during harvesting and threshing of seeds. The most intensive injuries reduce seed viability immediately, whereas, the small injuries may serve as an infection centre for the microorganisms. Generally, the large seeds are more susceptible to mechanical injuries than the smaller ones. Similarly, the spherical seeds are better protected against such injuries than the elongated and flat seeds.

b. **Environmental conditions at harvesting time:** A warm and dry weather at the time of seed maturity usually favour the production of quality seeds, whereas, adverse weather conditions like occurrence of frost, excessive rainfall or even drought at the time of harvesting of seeds may affect their viability adversely. Similarly, the temperature during seed ripening can also influence seed viability greatly. Nutrient condition of the plant at harvesting time is equally important. If the mother plants are nutrient deficient, they would produce seeds with low vigour, affecting their viability adversely. Likewise, the state of seed maturity at the time of harvest is important. For example, immature seeds, in general, are less viable than the mature seeds. Further, harvesting and extraction of seeds before full maturity is also detrimental to seed viability.

c. **Genetical factors:** Seed viability is also affected by genetic make up of the plant, as a result, there is variation in seed viability among different species by which these are produced. For example, seeds (stones) of mango have poor viability but it is not a problem in cashewnut, although both fruits belong to same family i.e. Anacardiaceae. Indian lotus has been reported to remain viable for 100 years.

d. **Storage conditions:** Proper storage conditions are important for keeping the seeds in good condition. If temperature, relative humidity, aeration and other conditions are not adequate in storage, it would surely affect the viability of seeds. As stated earlier, the recalcitrant seeds loose their viability under dry storage but the orthodox seeds require low moisture and cooler storage conditions for maintaining their viability. In general, high relative humidity shortens the life of the seeds. On the basis of moisture and temperature requirement of seeds, two rules should be kept in mind;

1. For every one per cent rise in seed moisture, the seed viability is reduced by 50 per cent. However, it is true only if the seed moisture is between 5-14 per cent.
2. For every 5°C increase in temperature, the life of the seed is reduced to 50 per cent. However, this range is applicable for temperature range between 0 and 50°C.

In addition to temperature and humidity, aeration is another potent factor, which determines the longevity of seeds. It means that there should be proper concentration of O_2 and CO_2 inside the storage house, or storage structure. If all these conditions are inadequate, these may favour the development of pathogens and storage insect-pests, which further worsen the condition of seeds.

Methods of testing seed viability

Before sowing, it is necessary to test the viability of seeds. To test the viability, a representative sample of seeds is taken from a lot and one of the following methods may be used for testing the viability.

1. Germination test

This is the most useful method of testing seed viability of seeds having no dormancy. In this method seeds are placed on moist blotting papers kept in petridishes in a refrigerator having the necessary optimum environmental conditions, *viz.*, temperature, light and humidity. The seeds may be kept in germinator, depending upon species, for 5 to 20 days. The development of normal seedlings from seeds is generally considered as germination. After the prescribed period, the final count of normal seedlings gives germination per cent or seed viability per cent.

2. Excised embryo test

This test is used to test the seed viability of woody shrubs and trees whose dormant embryo requires a long period for after ripening before germination. Usually seeds of such species do not respond to germination test. In such cases, excised embryo test gives fruitful and quick results. In this method, the embryos are excised from the seeds and soaked for one to four days by one of the following methods until they are completely swollen, (a) in slowly running water, (b) in standard water below 15°C, or (c) in standard water at 25°C with at least two changes of water daily. The soaked seeds are carefully placed in moist papers in petri dishes, kept in a room maintained at 18-22°C, normal relative humidity and light intensity. The embryos showing any growth activity (development of cotyledons, radicle and plumule etc) are considered as viable otherwise non-viable. The non-viable embryos become soft, brown and are infested by microorganisms very soon. The rapidity and degree of development of viable seeds give some indication of seed vigour also.

3. Tetrazolium test

This test is useful for early prediction of seed viability. In this method the seeds are first soaked in water, imbibed seeds are dissected longitudinally into two halves, exposing their embryos. These seeds are then placed in 1 per cent aqueous solution of 2,3,5-triphenyl tetrazolium chloride at room temperature for few hours to few days, depending upon species. The pH of the solution is adjusted in between 6-7 as the reaction goes smoothly at neutral pH. This chemical is highly sensitive to light and therefore any light exposure should be avoided. The exposed embryos of seeds will stain with the dye and the complete staining of embryos indicates full viability, completely unstained embryos indicate complete non-viability of the seeds and if the seeds are partially stained, it is very difficult to draw any conclusion.

4. X-ray analysis

The X-ray photographs show the emptiness of seeds, shaken embryos, number of embryos in seed and damaged portion of an embryo. If the embryo is fully developed and is un-damaged, the seed is said to be viable.

5. Miscellaneous methods

Electrical conductivity as a measure of planting seed quality is gradually catching the interest of workers as a measure of seed viability. The electrical conductivity test measures the amount of electrolytes, which leach from seeds during imbibition. The increase in conductivity has been found to be correlated with the decrease in germination and seed vigour in several crop species. When seeds are set to germinate, the imbibition of water is accompanied by a leaching of substances out of the seeds. These substances are mainly potassium, phosphates, sugars and amino acids. On the basis of correlation between seed deterioration and leakage of electrolytes from seeds, it has been hypothesized that the increase of electrical conductivity is due to the loss of ability to re-organize cellular membranes completely and rapidly during early imbibition.

An improved fluorescein diacetate (FDA) biochemical viability test has also been developed for viability test in gymnospermic seeds of plants such as jack pine, black spruce and white spruce. The efficiency of FDA test can be improved by slicing the seed coats prior to soaking in the stain, increasing the FDA stain concentration from 0.2% to 2.0% and adding a 45 minute incubation period (after the FDA stain soak) to the procedure. The amount of extractable fluorescein in seeds incubated with FDA solution is highly correlated with the proportion of live seeds.

Using computerized imaging technology, Seed Vigor Imaging System (SVIS) was developed at The Ohio State University, USA. By analyzing the digital images of seedlings, it can detect and measure the length of hypocotyls and radicles separately, and then generate a value for the growth and uniformity of a germinating seed lot. This system provides a fast, labor-saving and objective approach for measuring seed quality.

Another technology, known as dynamic speckle, or biospeckle, an optical phenomenon produced when active materials, such as biological tissue, are illuminated by laser light. Using this technology, the biological activity of seed tissues can be inferred from quantitative and qualitative measurements of their speckle activity. Similar to the tetrazolium test, the dynamic speckle technique aims to generate a map of the seed, identifying areas with different activities.

Seed vigour

Seed vigour is an important quality parameter, which needs to be assessed to supplement germination and viability tests to gain insight into the performance of a seed lot in the field or in storage. As per definition adopted by International Seed Testing Association Congress, (1977), seed vigour is the sum total of those properties of seed, which determine the potential level of activity and performance of seed or seed lot during germination and seedling emergence. Several factors like genetic constitution, environment and nutrition of mother plant, maturity at harvest, seed weight and size, mechanical integrity, deterioration and ageing and pathogens are known to influence seed vigour.

Measurement of seed vigour

The constraints imposed on vigour testing have limited the types of vigour tests, which can be employed for testing vigour of the seeds. Despite these limitations, several vigour tests have been developed, which has been described briefly hereunder:

1. Accelerated Ageing Test (AAT)

The accelerated ageing test has been developed at the Seed Technology Laboratory, Mississippi State University, USA for determining the storage potential of seed lots. In this method, the unimbibed seeds are subjected to high temperature (41^{0}C) and to cent percent relative humidity for 3-4 days. After removing seeds, these are allowed to germinate under optimum conditions. The relationship between initial moisture content and the seed germination is established to interpret the results. The seed lots that show high germination in accelerated ageing test are expected to maintain high viability during ambient storage as well. Therefore, ageing test gives an indication of the performance of the seed lot during ambient storage.

2. Cold Test

In this method, seeds are placed in soil or paper towel lined with soil and exposed to cold for a specific period to provide stress from imbibition, cold temperature and microorganisms. Afterwards, the seeds are placed under favourable growth conditions. The test aims to differentiate between weak and vigorous seed lots by subjecting them to low temperature prior to germination at optimum temperature. However, this has been criticized for using field soil, which greatly varies from place-to-place with regards to soils moisture content, pH, texture, structure, population of microorganisms etc., which may contribute to divergence in the results.

3. Conductivity Test (CT)

It is based on the known fact that weakening of cell membrane in poor vigour seeds causes leakage of water soluble compounds like sugars, amino acids, electrolytes etc. when immersed in water. On the other hand, fresh seeds having intact membrane leach less quantity of these chemicals. The measurement of electrical conductivity (EC) of the leachate by a good and sensitive conductivity meter gives an accurate estimation of membrane permeability. This method is very useful for measuring seed vigour of bulk samples. It is considered as a rapid, simple and inexpensive method; though, the initial seed moisture content and seed size should be standardized, which may affect the results.

4. Cool Germination Test (CGT)

This method of testing seed vigour is different to cold test in that 'CGT is conducted' under standard laboratory conditions at lower temperature (18.6 ^{0}C) and does not rely upon microorganisms for creating stress conditions in the germinated seeds. It is almost similar to germination test (GT) as its procedures are almost similar to the ones used for germination test. It is mainly used for testing the vigour of cotton seed.

5. Seed Growth Rate Test (SGRT)

This test is based on the concept that vigorous seeds have the ability to synthesize and transfer new materials to emerging embryonic axis more efficiently and rapidly, which results in accumulation of higher dry weight in the seedlings. Thus, the vigour results are expressed as dry weight/germineable seedling. In this method, the growing segments of the embryo of normal seedlings are excised from the storage organs, like cotyledons or endosperm, placed into beakers and dried at 80^{0}C for 24 hours and then increase in dry weight is measured. This test is also similar to the routine germination test. However, the results of this method are mostly precise, because the seedling growth rate is correlated with vegetative development of the seedling in the field. The differences in moisture and light intensity can influence the results, which is considered as the major limitation of this test.

6. Seedling Vigour Classification Test (SVCT)

This is also similar to routine germination test except that normal seedlings are further classified into 'strong' and 'weak' categories. In general, it is not considered more reliable test because the evaluation of the seedlings is done subjectively.

7. Tetrazolium (TZ) test

It is one of he most reliable tools for seed analysis, which has been described in detail under texts for seed testing/viability.

8. Hiltner Test/ Brick gravel test

The test got its name on its developer, Hiltner in Germany in 1917. This was developed for screening out the *Fusarium* infected weak seeds in cereal crops. Hintler observed that the seeds of cereal crops affected by Fusarium disease were able to germinate in regular test but were not able to emerge from brick gravels of 2-3 mm size. On the other hand, healthy seeds were able to emerge from the brick gravel. The principle is that the weak seedlings are not able to generate enough force to overcome the pressure of brick gravels, so this method can be used to differentiate vigour levels in cereal seeds.

* * *

3 Vegetative Propagation

Many horticultural and forestry plants don't come true-to-type when propagated by seed owing to cross-pollination. Due to cross pollination fruit plants are highly heterogeneous nature of fruit crops, in general. As a result, when such plants are grown, the individuals differ in morphological as well as fruit characteristics as compared to mother plant. Though, it was a common practice in old days to establish new orchards with seedling plants, but now with the development and standardization of different vegetative propagation techniques, such plantings are avoided. In modern system of crop production, instead of a single tree derived from seed, use of composite plants (consist of usually scion and rootstock, joined together by budding or grafting) is in vogue. This could have been possible by growing interest of mankind in vegetative means of plant propagation. In asexual propagation, vegetative parts of the plant are used for propagation. It is practicable because all the living cells of a plant have a capacity to regenerate into a full plant under favourable environmental conditions. The plants produced through vegetative means are therefore genetically identical and similar to the mother plant, possessing all the characteristics of the mother plant. Vegetative methods of propagation are important because they permit large-scale multiplication of a genetically identical individual plant. Vegetative methods of propagation have certain advantages and limitations over sexual (seed) method of propagation, as given below;

Advantages

In some respects, different methods of vegetative propagation are advantageous over sexual methods of propagation in several ways, which are discussed as under:

1. Vegetatively propagated plants are true-to-type and are identical to mother plant.
2. Vegetatively propagated plants have uniform growth, and give consistent yield of better fruit quality.
3. Fruit trees come into bearing earlier than the seedling trees.
4. Vegetative propagation is the only choice in plants, the seeds of which are more prone to dormancy (peach and olive etc.).

5. Some fruit plants like banana, pineapple, fig and varieties of grape, guava and lemon have seedless fruit and produce little or no viable seeds. Such plants are easily perpetuated through vegetative means.
6. Vegetatively propagated plants are less vigorous and thus their management is easy and less costly than seedling plants.
7. With vegetative means, a large plant is obtained within short time.
8. The desirable abiotic effects of rootstocks on the scion cultivar can be utilized effectively by budding or grafting them on desirable rootstocks. Regulation of the tree size, fruit quality, precocity etc. by using different rootstocks or interstock is possible.
9. A 'clone' can only be perpetuated through vegetative means.
10. Self-incompatibility exists in a number of fruit plants like mango, aonla, apple, almond and cherry etc. Provision for adequate cross-pollination in an established fruit orchard can be obtained by top working in scattered trees throughout the orchard with the proper pollinising variety.
11. Indicator plants can be used for detecting various viral diseases through viral indexing technique in which the desirable scion is grafted or budded on to a susceptible rootstock. For example Kagzi lime can be used as rootstock (indicator plant) for detecting the presence of Tristeza virus in citrus.
12. Early assessment of hybrids/varieties can be done by grafting or budding the desirable hybrid/variety on a large established tree or on certain dwarfing rootstock.
13. Damaged part of the tree trunk or root can be repaired by bridge grafting or inarching.
14. A number of varieties can be grown on a single plant, which is a novelty.

Limitations

Vegetative propagation has certain limitations as under:

1. The vegetatively propagated plants, particularly budded and grafted are short-lived.
2. No new variety can be evolved by vegetative mean of propagation.
3. It is a specialized task and requires technical skill.
4. Vegetatively propagated plants are usually more susceptible to insect-pests and diseases.
5. Vegetatively propagated plants are usually prone to suckering.
6. Vegetatively propagated plants are usually less hardy to abiotic stresses as compared the plants raised through seeds.
7. Vegetative propagation is more expensive than the seed propagation.

8. Vegetative propagation is failure in some fruit plants and forestry trees in which no method of vegetative propagation has been found successful. For instance, papaya and phalsa.

A large number of horticultural and forestry plants are raised through one or the other vegetative methods. Many methods of vegetative propagation have been standardized for different crops. However, one method may be highly suitable for a particular crop but it may be completely failure for the other. Similarly, one crop may be propagated on a large scale with different methods of propagation, whereas, the other may have only one method. The success of different methods is influenced by the environment also and as result, one method for a particular crop may be highly successful in a particular locality, but can't be of any value in the other. For example, side veneer grafting is highly successful method of mango propagation in north India but in *Konkan* region, epicotyl or stone grafting is only successful. A list of general vegetative methods employed for raising different horticultural and forestry plants has been given below:

Methods of vegetative propagation

1. *Propagation by apomictic seedlings*

   (Apple, citrus etc.)

2. *Propagation on its own root system* Propagation by cuttings

   (hard-wood cuttings, root cuttings, leaf cuttings etc.) Propagation by layering

   (Tip layering, ground layering, mound layering, air layering etc.)

3. Propagation on the root system of other plants

   (a) Propagation by grafting

   (Splice grafting, tongue grafting, approach grafting, stone grafting, veneer grafting, side grafting etc.)

   (b) Propagation by budding

   (T budding, inverted 'T' budding, patch budding, ring budding, etc.)

4. *Propagation by specialized vegetative structures* Propagation by bulbs, Propagation by tubers, Propagation by rhizomes, Propagation by corms, Propagation by suckers, Propagation by runners, Propagation by offsets, Propagation by bulbils, Propagation by pseudobulbs, Propagation by stolon, Propagation by other vegetative structures (slips, crown etc.)

5. Micropropagation

Propagation from the minute plant parts under highly aseptic and controlled conditions of environment is called as micropropagation. In micropropagation, different explants (shoot tip, embryos, cell or protoplast etc.) are used. It has been described in detail in a separate chapter.

Vegetative propagation, in general, is very useful for the multiplication of heterozygous plants, which don't breed truly by seeds. The details on principles and methods of the above mentioned vegetative methods are given in the subsequent chapters.

Polyembryony and apomixis

Occurrence of unique phenomenon of polyembryony and apomixes in some fruit and forestry plants represent variations from natural pattern of zygote formation and enbryogenesis. Their occurrence can develop from many reasons and causes as discussed briefly hereunder:

Apomixis

Apomixis is asexual reproduction through seeds, i.e., embryos that develop within seeds arise from cells of the maternal genotype and are not products of genetic recombination and syngamy (fusion of male and female gametes). The seedling plants raised through such seeds are called as *apomictic seedlings.* The genetic make up of apomictic seedlings is similar to the mother plant and hence propagation by means of apomictic seedlings is similar to vegetative propagation. Plants that produce only apomictic embryos are called *obligate apomicts* and the ones, which produce both apomictic and sexual embryos, are called as *facultative apomicts.* Apomictic plants are the result of either one of two types of development, sporophytic or gametophytic and on this basis, apomixis may of the following types:

1. **Diplosporous apomixis:** In diplospory, the megaspore mother cell by passes or fails to achieve meiosis, but through mitosis, forms an embryo sac with all unreduced cells distributed as in the meiotic embryo sac of the *Polygonum* type. In this case, the sexual process is completely compromised. The reductional meiosis is replaced by a restitutional meiosis or by mitosis.
2. **Aposporous apomixis:** In apospory, some nucellar cells, called aposporous initials enter in mitosis directly and unreduced embryo sacs are formed. Sexuality and apospory can occur simultaneously in the same ovule. If fertilization of the central cell is necessary to form the endosperm, the system is regarded as pseudogamous and if not, it is autonomous. The reduced megaspore degenerates and is replaced by a somatic cell. Both diplospory and apospory are forms of gametophytic apomixis, in which a megagametophyte or embryo sac produces a parthenogenetic embryo. Apospory has been noted in genus *Malus* and *Rubus.*
3. **Adventitious embryony:** The phenomenon of polyembryony was discovered by A. Leeuwenhoek in 1719, who found two germlings developing from the same

seed in *Citrus*. Later, Strasburger (1878) described the formation of adventive embryos from nucellus cells in *Citrus*. He called this phenomenon nucellar or adventive polyembryony. In sporophytic apomixis or adventitious apomixis, the embryo does not develop from the cells of the embryo sac but from a cell or a group of cells either of nucleolus or integuments and is diploid, and hence the resultant embryo is also diploid. Such embryos usually develop outside the embryo sac in addition to the regular embryo. This phenomenon is quite common in citrus and mango, where pollination and fertilization takes place in usual manner and both zygotic and apomictic embryos are produced together. In some species, such phenomenon does not need any pollination and fertilization as in *Opuntia* spp. The comparision of different apomicts with the sexual seed production and adventitious embryony has been shown in **Figure 7**.

4. **Vegetative apomixis:** In this type, vegetative buds or bulbils are produced in the inflorescence in place of flowers. These buds or bulbils may sprout into new plants while they are still on the mother plants. This type of apomixis is quite common in *Allium, Agave, Poa* and *Dioscorea* etc.

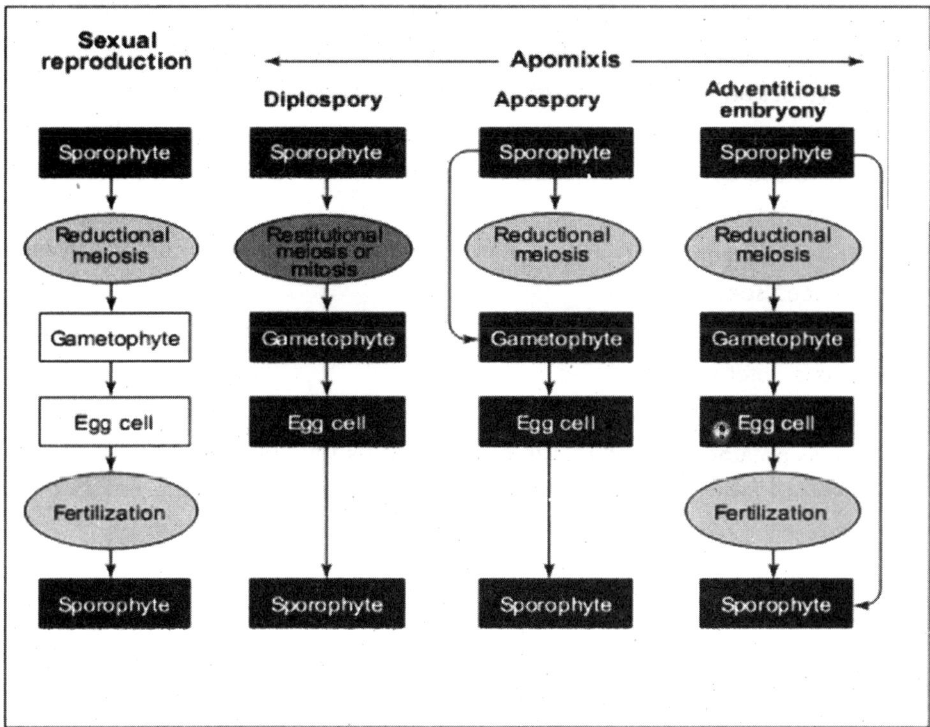

Fig. 7. Comparison of sexual and asexual reproduction. Reduced life cycle stages (n) are in white, unreduced stages (2n) are in black (Adapted from Dijk and Damme, 2000).

Significance of apomixis

Apomixis occurs in many plant species in nature. The ability of plants to reproduce apomictically has great significance in horticulture because of the following reasons;

- Apomixis provides an opportunity for producing genetically uniform population.
- A major potential of apomixis is that it could lead to "hybrid crops that clone themselves". Fixing and cloning through seeds elite genotypes and hybrids are achievable through apomixis. This would, in turn, allow small farmers to propagate their own seeds. As a result, commercial production will be simplified, with a subsequent decrease in cost. The transference of apomixis gene(s) in a breeding program was successfully performed in guinea grass (*Panicum maximum* Jacq.). "Natsukaze" was the first apomictic hybrid cultivar generated from crosses between a sexual plant and an unidentified tetraploid apomictic line through open pollination in Japan.
- Substantial increase in yield in some crops due to photosynthate availability through elimination of male flowers/flower parts.
- Apomixis helps to avoid the complications associated with sexual reproduction, such as pollinators and cross-compatibility.
- Apomictic seedlings are useful for producing uniform rootstocks in apple (*Malus toringoides* and *M. sikkimensis*), citrus and *gamun* etc.
- Polyembryony demonstrates that embryogenic potential is not limited to zygote but is possessed by various other somatic cells, which forms the base for *in vitro* somatic embryogenesis.
- Apomictic seedlings are highly uniform and quite healthy in contrast to sexually produced seedlings.
- Apomictic seedlings are helpful for identification during hybridization (e.g. trifoliate leaf character in citrus).
- The avoidance of viral transfer in plants that are typically propagated vegetatively, such as potatoes, can be realized by harnessing the potential of apomixis.

* * *

4 Selection and Management of a Clone

The biological process of producing plant similar to source/mother plants is known as *cloning* and the resultant population is called as *clone*. In nature the process of cloning is achieved by asexual means e.g. through production of vegetative structures like bulbs, runners, rhizomes etc., by plants. Vegetative propagation renders an opportunity to select, maintain and propagate a desirable type indefinitely.

Origin of a clonal cultivar

Most of the clonal cultivars have originated either through seedlings, mutations or through biotechnology.

Seedling selection: Most cultivars grown in different parts of the world have originated from a single plant in a seedling population. This original population might have either originated as a chance seedling or through careful planned breeding programme. Clonal multiplication of the selected types is either done by (1) multiplication from the same plant, and/or, (2) consecutive generations of the progeny from the same plant by the breeder. In general, introduction of a seedling selection in a particular area/locality is done by conducting a series of experiments because genotype x environmental interactions have great bearing on its performance in that particular locality. This is the reason that one variety may excel others in a particular area, but may be completely failure in the other.

Many cultivars of horticultural importance have been selected as chance seedlings. For example, most cultivars of mango grown in India like Dashehari, Langra, Chausa, Bombay Green, Baneshan, Neelum, Alphonso etc., have been slected as chance seedlings. The world famous Bartlett pear, Delicious apple, Gros Michel banana, Sultana grape etc., have also been selected as chance seedlings. Similarly, some ornamental shrubs, woody perennials are also clonal selections.

Mutations: Chance mutations within established clones are the second major source of new cultivars. A mutation is a sudden genetic change involving some part of DNA molecule.

It may result in structural changes in DNA molecule i.e., a change in chromosome number. Further, mutation can be brought out artificially with the treatment of mutagenic agents like, X-rays, gamma rays and certain chemicals like colchicine. Striking examples of natural mutations have appeared in apple, grape and citrus. In these fruit crops many cultivars are bud sports, which have suddenly appeared as a change in the branch of a plant, which later proved better than mother plant e.g. Rubired grapefruit is a bud sport.

Biotechnology: It is a modern science in which a number of strategies can be used to alter the genotype of a clone. Recombinant DNA technology in which DNA fragments of one organism are cloned and transferred to another clone. In horticulture and forestry, this technique has great significance because a single and desirable trait can be transferred in to an established cultivar, lacking in that particular desirable trait.

Production and maintenance of a clone

The production of a clone needs to involve the procedures that maintain the genetic purity, detect modifications and exclude pathogens. It is a known fact that clones may deteriorate owing to these factors. It is likely that a clone does well in a particular locality but may fail in the other owing to different environmental factors or otherwise.

Variability within a clone

In general, the plants of a clone are genetically pure but the actual appearance and behavoiur of the plants of a clone can be changed with the environment in which it is growing. For example, Thompson Seedless cultivar of grape produce crop of excellent quality in Maharashtra and South India but does not perform well under north Indian conditions. Similarly, Dashehari, the commercial mango cultivar of north India and Alphonso, the commercial cultivar of western India, don't perform very well when grown in western and north Indian conditions, respectively. The environmental factors, however, don't change the genotype of the plants. The fruit trees of the same cultivar in an orchard may even differ because of difference in soil, water availability or a rootstock. The life of a clone is unlimited but may deteriorate if it is continuously exposed to unfavourable environment. Further, pathogens, especially viruses or virus like organisms, may infect clone and can cause deterioration in it. Similarly, genetic changes brought about by mutations can produce "off type" individuals and reduce the value of the clone. However, mutations in certain cases have even been found to be useful in the production of superior genotypes also. The notable examples are Low Gate banana from High Gate cultivar; Sonaka, Manik Chaman grapes from Thompson Seedless grape, Tas-e-Ganesh grape mutant from Anab-e-Shahi cultivar and Dashehari-51 clone from Dashehari mango. Basically, the variability between juvenile to adult growth and developmental phases brings the difference in the plant. The plant in its life undergoes three different growth phases, designated as juvenile phase, transition phase and adult phase. These phase changes have great significance in plant propagation. Plant propagators must recognize the variability factors associated with these changes and measures to overcome them. With the phase change phenomenon, generally two significant aspects of propagation are associated. Out of these, one may be that the juvenile, transition and adult phases occur simultaneously in different parts of a plant as in apple, pear, citrus in which the flowering begins in the outer and upper portion of a plant after some years of juvenility. The second aspect is that the vegetative propagation of a meristem reproduces physiological and morphological features of the particular phase that it had on the mother plant and it can show much variation both among the plant produced, mother plant and the offspring. Thus, it is evident that plants propagated from the buds taken from adult phase flower early than the ones taken from the juvenile phase.

Genetic variations within a clone

In general, individuals within a clone are usually considered to be the true copy of the mother plant. However, certain variations may occur in them in due course of time. These genetic variations within a clone may take place through mutations, chimeras, transpsons, and somaclones as described briefly hereunder;

A. Mutations

The term mutation was first coined by De Varies in 1905, who designated mutations as the sudden heritable changes occurring in the cells of an individual, which may bring permanent genetic changes in a part of a clone. A plant that arises as a result of mutation, is known as *mutant.* Mutations may occur due to deletion, duplication or inversions of some parts of chromosomes or due to addition or deletion of the chromosomes. In some cases, changes may be brought about by changes in the DNA of plastids (plastone) and mitochondria (chondrianome), which may results in the development albino (no chlorophyll) or variegated plants. Mutations may occur spontaneously or induced artificially through the use of mutagens. Its rate primarily depends upon the location of the cell in the meristem tissue. Cells in the apical meristem are more stable and are likely to mutate less. The changes usually go unnoticed until and unless they occupy a large part or an entire branch of a plant such as a bud sport. The changes in colour (variegation) can, however, are detected easily. Usually, the results of mutations are defective and are more often inferior as in citrus. However, some mutants have been found useful for selecting new clones of superior varieties in horticultural crops. For instance, pink fleshed grapefruit, a superior type is a bud sport of white fleshed grapefruit. Washington Navel orange is a bud mutant of Laranga Selecta Brazilian orange. Dashehari-51 is a prolific and regular bearing mutant of Dashehari mango.

B. Chimeras

A plant is said to be a *chimera* when cells of more than one genotype (genetic makeup) are found growing adjacent in the tissues of that plant. Therefore, the development of plants with two or more genetically different tissues growing separately but adjacent to each other, usually called as *chimera*. Chimeral plants may originate by grafting, spontaneous mutation, induced mutation, sorting-out from variegated seedlings, mixed callus cultures, or protoplast fusion. Colchicine has been widely used to induce *cyto-chimeras* of fruiting plants. Variegated plants are perhaps the most common types of chimeras, and certainly the most convenient example to use in presenting the basic concept. All the cells in a variegated leaf originate in the apical meristem of the shoot, but some cells are characterized by the inability to synthesize chlorophyll. These appear white rather than green even though they are components of the same tissue system. Many important selections of foliage, floricultural, and landscape plants are chimeras. *Cornus alba* 'Argenteo Marginata', *Vinca minor* 'Variegata', *Ajuga reptans* 'Burgundy Glow', and many selections of *Hosta, Diffenbachia, Peperomia, Chlorophytum,* and *Saintpaulia* are few examples of the chimeras, which have variegated foliage. When a mutation occurs within a single cell of a clone, it initially produces an island of mutant cells and the plant becomes a mixture of two different genotypes and if the mutated tissues grow successfully, it results in variegation. Further, mixture of sour and sweet type apple, fuzzy and smooth peaches and thornless blackberries are some other examples of chimeras. Similarly, occurrence of red colour in outer skin of apple and potato with mutated inner layers are also some examples of chimeras.

Tunica-Corpus Model of Apical Organization

This model helps us to understand the development of different kind of chimeras in plants. The apical meristem of a shoot is the location where most of the cells, which produce the plant body, are formed. The apical organization of most dicots follows the tunica-corpus pattern. According to this concept, the meristematic region above the youngest leaf primordium is organized into two zones of cells, a layered region (the tunica) and a region where layering is not evident (the corpus). The controlled pattern of cell divisions in the tunica results in the maintenance of discrete layers or *histogens*. The tunica layers are usually identified as L-I (outer), L-II (middle), L-III (inner) and so on (Figure 8).

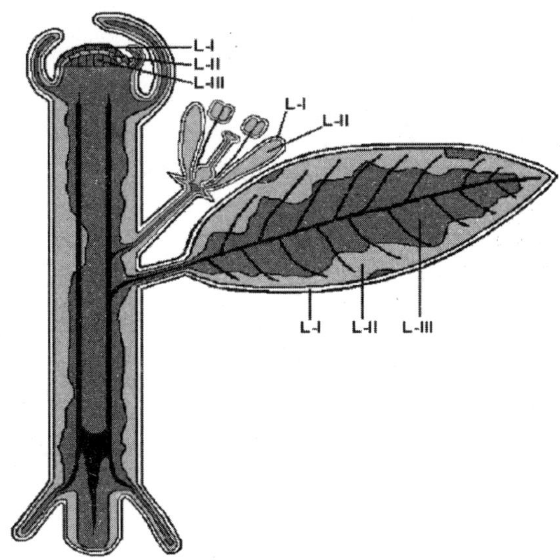

Fig. 8. Depiction of different tunica layers in plant system.

The derivatives of the different layers are discussed in **Table 2.**

Table 2. Derivatives of different tunica layers.

Layer	Derivatives
L-I	Epidermis of all organs. Monocot leaves: L-I contributes to the outermost region of the leaf mesophyll giving rise to a strip along the leaf margin. Dicot leaves: L-I usually gives rise to only the colorless epidermis, thus cannot be seen; sometimes L-I gives rise to small islands of tissue along the margin.
L-II	Stem and roots: Outer and inner cortex and some of vascular cylinder Leaves: Mesophyll in outer region of leaf
L-III	Stem and roots: inner cortex, vascular cylinder and pith Leaves: Mesophyll in central region of leaf

The inner zone, or corpus, is not layered as is the tunica. Chimeras arise when a cell undergoes mutation. If the cell, which mutates, is located near the crest of the apical dome, then all other cells which are produced by division from it will also be the mutated type. The result will be cells of different genotypes growing adjacent in a plant tissue giving rise to a chimera. Structural classification of chimeras includes periclinal, mericlinal and sectorial chimeras (**Figure 9**). Periclinal describes the stable, "hand-in-glove" arrangement of the tunica-corpus region; mericlinal, describes a type of periclinal where only part of a layer is mutant, and sectorial, describes a form where a solid sector through all apical layers is mutant.

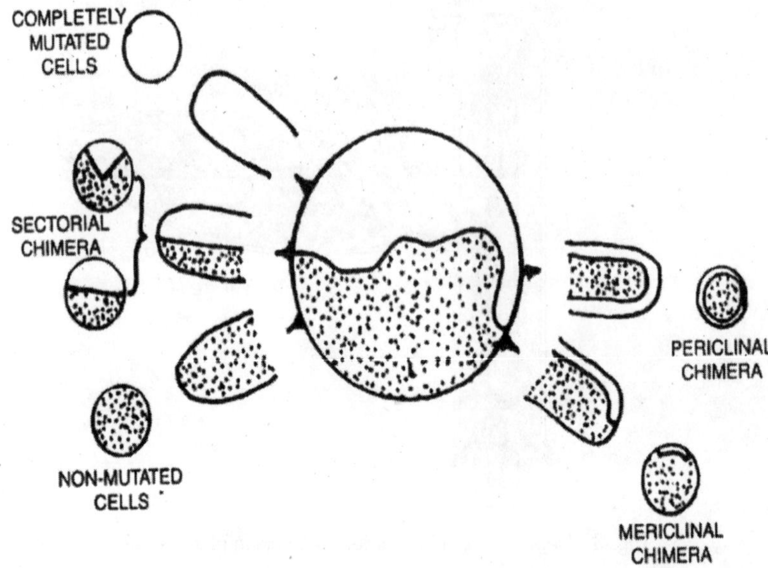

Fig. 9. Different kinds of chimeras encountered in nature.

Types of chimeras

1. Periclinal chimera: Periclinal chimeras are the most important category since they are relatively stable and can be vegetatively propagated. In this chimera, a tissue with one genotype occurs comparatively as a thin layer around genetically different core (non-mutated tissue). This is a relatively stable form of chimera if propagated through stem or grafting. However, it is not stable, if propagated by other methods, such as, root cuttings. A typical example of this type of chimera is the genus *Rubus,* which has many thornless blackberry types. The epidermal layers in these types lack the gene for thorniness. If such types are propagated through stem cuttings or tip layering, the new plant will be thornless. However, it may be significant to note that the roots of such thornless plants arise endogenously from the non-mutated tissues and if such roots are used for subsequent propagation, the plants will be thorny. Similarly, if seeds of thornless plants are used for propagation, the seedling plants will be thorny, because the gametes are produced from the cells of the tissues, which originated from non-mutated layer. However, rarely some thornless forms may also be produced. Similarly, in some red coloured apples, the red pigment is located only in epidermal layers and the inner layers are green or yellow in colour. Propagation of periclinal chimera through leaf cuttings may, sometimes, results in non-mutated forms. For example, if variegated forms of *Sansevieria* or *Peperomia* are propagated through leaves, it is possible that the new plants are green or albino (**Figure 10**).

Fig. 10. Adventitious shoot formation on leaf cuttings of a periclinal chimera of *Peperomia*.

2. Mericlinal chimera: In this type of chimera, the combination is similar to periclinal chimera except that the mutated tissue occupies only a part of the outer layer and does not extend completely around the shoot as in periclinal chimera. It occurs more frequently because it is the resultant of mutation of a single or a few cells. It is not a stable but with continued vegetative propagation, it may either convert to periclinal chimera or to non-mutated tissue or a non-chimeral shoot. For example, a lateral bud arising from a mericlinal chimera would become a periclinal chimera (shoot) and from a normal part to develop in to a normal shoot. Further, if propagation is done from the terminal growing point or from lateral shoots at the tip of mutated portion, the resultant plant will be mutated one.

3. Sectorial chimera: In this type, the mutated cells occupy an entire sector of the stem including all layers of the shoot apex and the shoot may be composed of two genetically different tissues, occupying two distinct sectors in a shoot. The leaves or buds arising from such chimeral shoots will also be of two different identities. These chimeras appear if mutations occur in roots or in embryos at a very early stage where the cells of growing point don't occur in layers. These types of chimeras are, however, common in roots than in shoots. It is also an unstable form of chimera but with continuous vegetative propagation, the plant develops, may either be normal, mutated or periclinal chimeras. Sectorial chimeras may sometimes result in the development of variegated plants. The variegation in plants is caused by mutations on plasma genes, which affect the chlorophyll development. Due to this, some parts of the leaves produce green-white (albino) colour patterns. However, variegation may sometimes be caused by the attack of viruses. For example, colour breaks in some plants (tulips) are caused by viruses rather than by mutation. Likewise, some variations are caused by "pattern" genes that continue to reproduce that pattern.

4. Graft chimeras: Although, most chimeras occur naturally, however, these can be induced artificially by grafting. Basically, the concept of chimera came into existence in 1644 with the development of "Bizzaria" orange in Florence, Italy, following the grafting of sour orange (*Citrus aurantium*) on to citron (*C. medica*) rootstock, which led to the development of word chimera. Later, Winkler (1910), a world famous scientist, has produced numerous graft chimeras of tomato many years ago by grafting tomato on *Solanum nigrum* and *vice versa*. In the grafting process, the scion of the grafted plant was cut severely near to callus, which arises during the healing process. Adventitious shoots, which arise from the callus, are chimeras of tomato and nightshade (**Figure 11**), the name given by Winkler (1910) after the mythological monster that was part lion and part dragon. Likewise, an intergeneric graft chimera between *Citrus reticulata* 'Ponkan' and *Poncirus trifoliala* L. Raf. was found in a farmholding orchard in Fujian Province, in 1986, which was obtained from the buds appeared after freezing damage in graft union of 2-year-old seedling of cultivar 'Ponkan' grafted on the stock of *Poncirus*. The graft chimera is named as +*Citroponcirus* 'Hormish' which has unifoliate, bifoliate and trifoliate leaves in some branches, and trifoliate in other branches, showing a chimera of the two species.

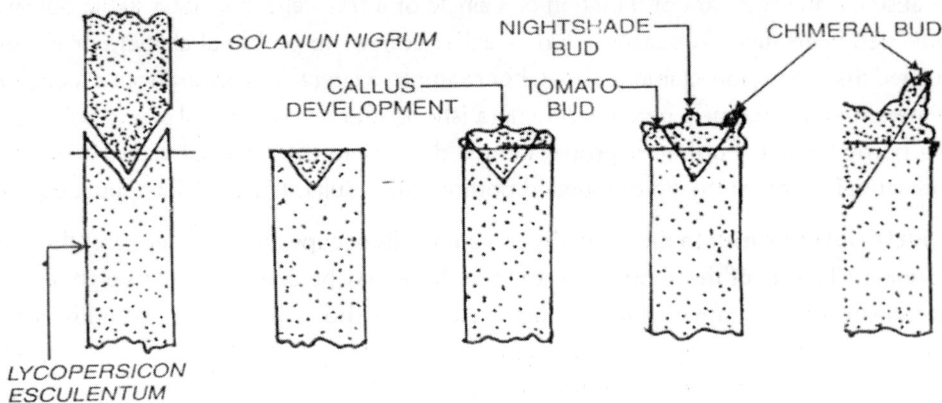

Fig. 11. Various stages of development of a graft chimera in the nightshade and tomato graftage.

C. Transposons

Genetic variation in some plants may take place due to transposable elements also. This variation typically involves pattern of coloured structures and dots of different sizes and arrangements. For example, variegated seeds of Indian corn, veriegation in poinsettia, and sansvieria. Transposons are defined as sequences of DNA, which can move or transpose themselves to new positions within the genome of a single cell. The mechanism of transposition can be either "copy and paste" or "cut and paste". Transposition can create phenotypically significant mutations and alter the cell's genome size. Such genes have also been called as "jumping genes". The first transposons were discovered in maize (*Zea mays*), by Barbara McClintock in 1948, for which she was awarded a Nobel Prize in 1983.

D. Somaclonal variations

Somaclonal variation is the term used to describe the variation seen in plants, which have been produced by plant tissue culture. Chromosomal rearrangements are an important source of this variation. In fact, this type of variation occurs in the cells in the deeper areas of growing point below meristem, which often become polyploid. There are both benefits and disadvantages of somaclonal variation. The major benefit of somaclonal variation is in plant improvement as somaclonal variation leads to the creation of additional genetic variability. On the other hand, a serious disadvantage of somaclonal variation occurs in operations, which require clonal uniformity, as in the horticulture and forestry industries where tissue culture is employed for rapid propagation of elite genotypes. Therefore, somaclonal variations have been regarded as a boon to the breeders but a curse to the propagators.

Non-genetic variations within a clone

Genetic variations within the clones usually lead to permanent changes in the plants. Such changes are sometimes desirable, and even may lead to the selection of new varieties. However, sometimes these may be undesirable and thus lead to the production of 'off type' plants, which fetches poor price in the markets. In addition to genetic variations, certain non-genetic changes may occur within the clones, which have been discussed here under;

i) **Environmental effects on clones:** Horticulturists of different parts of the world have one common view that though individuals of a clone are genetically similar but their performance is variable when plants of a single clone are grown at different sites and under different management practices. This is probably due to genotype x environment interaction, which may affect the phenotype (the appearance) or the performance of a clone.

ii) **Phase variation and change in maturation:** Phase variation is an important non-genetic variation. Its application can be beneficial to plant propagators depending upon the species and cultivar. Plants differ in their ability to change in phenotype with age. For example, those plants, which don't show much change in phenotype with age are called as *homoblastic* and those, which show distinct variation in specific traits with age are called as *heteroblastic.*

iii) **Topophysis:** Topophysis is the phenomenon in which different parts of a plant show phase variation and if propagated through meristems, the different phases are perpetuated in the offspring. This phenomenon is also called as determination of the meristems. Plants of certain species produced by cuttings taken from upright shoots (*orthotropic*) will reproduce plants in which the shoots grow vertically. Similarly, plants produced by cuttings taken from lateral shoots (*plagiotropic*) will grow horizontally. Many examples can be cited in which topophysis is manifested following vegetative propagation. For example, the new shoots from cuttings in coffee grow horizontally or laterally depending on the branch of the mother plant from which cuttings are taken. Some conifers and *Araucaria sp.* show strong tendency to manifest topophysis, when propagated through cuttings. The mechanism involved in the control of phase changes is not well understood; however, epigenetic variation and physiological changes may perhaps be associated.

* * *

5 Propagation by Cuttings: Principles and Practices

Adventitious root formation is a key step in vegetative propagation of woody or horticultural species, and it is a complex process known to be affected by multiple factors. Roots, which arise in the plant from sites other than their normal sites in the embryo or the primary roots, are defined as *adventitious*, as are roots that arise on explants from stem or leaf tissues cultured *in vitro*. In propagation by cutting, a portion of root, stem or leaf is detached from the mother plants and several chemical, environmental, mechanical treatments are given to induce the formation of root and shoot. A cutting, when put in suitable propagation medium becomes an individual plant in its own root, similar to the mother plant. Raising of plants through cuttings is a simple, quick, economical method of plant propagation and does not require the technical skill as necessary in other propagation methods like grafting, budding and micropropagation. Most of the plants are raised by stem cuttings, but in some cases plants can also be propagated from leaf and root cuttings. The plant part to be used as cutting chiefly depends upon the growing habit of the plant. However, any plant part taken for cutting must reproduce into full plant.

Histologically, the formation of adventitious roots within tissues, which usually do not produce roots, requires that a cell, or group of cells to begin some developmental process (such as competence, determination and meristemoid formation) to form roots. In general, the rooting competence of some cell types remains unexpressed *in planta*; however, it may be induced by culturing them *in vitro*. This could be attributed to wounding reaction, as such or in combination with the action of specific growth promoters/hormones, for instance, auxin. In genera, such as willow, poplar, jasmine and citron, preformed adventitious root primordia may already be present *in planta*. These preformed primordia normally lie dormant until the stem is made into cuttings and placed under conditions favourable for emergence as adventitious roots. Competence for root formation is defined as the ability of specific cells in a tissue to respond to specific root-inducing stimuli. Once competent cells have been exposed to a root inducer, they may become determined for root formation.

A key stage in adventitious rooting is the formation of root meristemoids (a clump of small cells growing through synchronous mitotic divisions). These root meristemoid later develops into root primordia. Subsequent primordium development leads to root emergence.

Advantages

i. It is the easiest and cheapest method of propagation.

ii. The plant grows on its own system, therefore complex problems of stock-scion relations, that exist in budded/grafted plants, can be avoided.

iii. Rapid multiplication is possible within a short time.

iv. Much space is not required.

v. Transportation of cuttings is easier.

Disadvantages

i. Cuttings of some plants do not root easily.

ii. The desirable benefit of rootstock on scion variety can not be exploited.

Rooting of cuttings

Induction of rooting is the primary requirement for propagation of plants through cuttings. The root formation from cut portion of cuttage is a genetically and physiologically governed phenomenon. Several intentional manipulations by propagators in terms of chemical, mechanical and environmental treatments are required for proper induction of roots from the cuttings. In principal, regeneration of a new plant from a cutting basically depends on two fundamental properties of the plant cell. First is totipotency, which states the individual cell contains all the genetic information required for producing a new plant of same kind, and the second is dedifferentiation, i.e., the capacity of mature cells to return to a meristematic condition and develop in to a new growing point. Thus, for transformation of a plant part in to a full-fledged plant, a plant propagator has to do some manipulations for inducing root and shoot system in the cuttings. In general, there exists greater variability amongst plant species in the ability to root in their cuttings. In some species, cuttings root very easily, as in grape, in some species, rooting takes place only after some treatments, while others do not root at all. Therefore, it becomes mandatory to study the anatomical and physiological basis involved in the process of rooting of cuttings.

A. Anatomical basis of rooting of cuttings

Anatomy of plant part from where the cutting has taken, plays a vital role in the process of root induction. The roots may be developed on stem, root or leaf cuttings but they all have internal origin in their parent structures. The roots to develop (adventitious roots), either in stem, root or leaf cuttings, have internal origin in parent structures of most plant species. However, in some species (e.g. *Tamarix*), they may develop exogenously and then connect to the internal tissues of the stem. In the process of rooting of cuttings, generally four anatomical changes are observed. These are, i) Formation of place for root initials by de-differentiation of specific mature cells, ii) Formation of root initials from certain cells near vascular bundles, iii) Development of root primordial and formation of vascular connection between root primordia and conducting tissues of cuttings and emerging through the cortex and epidermis, iv) Emergence of roots outside the cuttings.

The herbaceous stem has four major areas i) a large pith in centre, ii) a ring of vascular bundles outside the pith, iii) a cortex outside vascular bundles, and iv) a thin epidermis. In herbaceous plants, adventitious roots are formed just outside and between the vascular bundles, though it may vary depending upon kind of plant. For instance, roots arise in phloem parenchyma region in tomato, pumpkin, from epidermis in crassula and from vascular bundles in castor. In woody plants, one or more layers of xylem and phloem are present and roots are formed in stem cuttings from living parenchyma cells. The steps involved in rooting of cuttings have been described here under.

Callus formation and rooting: In the early part of nineteenth century, research work was initiated on how the actual rooting process is initiated in the cuttings. Later, it was observed that when cuttings are placed in a suitable medium, a mass of undifferentiated parenchymatous cells, called as *callus*, is usually developed at the base of the cuttings and only then the root initiation process takes place. Then, it led to belief that callus formation is necessary for rooting of cuttings, but now it has been established that callusing and root initiation are two independent phenomenon and can occur simultaneously. Though, the excess callus formation may hinder root initiation in some species, it may; however, be a precursor for root initiation in others as in *Sedum, Hedera helix* etc.

Presence of preformed root initials: Adventitious root initials are developed in the intact stem of certain woody plants even before cuttings are made. These root initials develop to some extent, become dormant and get embedded within the stem. Such dormant root initials are called as preformed or latent root initials. When the cuttings are made from such stems are placed in favourable environmental conditions, the dormant or latent root initials become active and roots are developed rapidly and easily. Occurrence of root initials is quite common in willow, currant, citron and old plants of apple, cherry, and quince.

Stem structure and rooting: It has been found that certain type of stem structure or tissue relationship plays a vital role in adventitious root formation in the cuttings. The presence of a continuous sclerenchyma ring, more exterior with respect to the point of origin of the root, possibly constitutes an anatomical barrier to root emergence, and such a ring has been associated with difficult-to-root cuttings, while easy-to-root types have been characterized by discontinuity in the sclerenchyma ring. Further, in some species, some structures within the stem favour initiation of root primordia than others. For example, citron (*C. medica*) produces root profusely from preformed root initials within a short time compared to sour orange (*C. aurantium*), which forms only a few roots that too after several weeks. The true problem of the absence of a rooting response, as is the case with several recalcitrant fruit species, remains the absence of meristemoid formation. However, some anatomical peculiarities may enhance the rooting possibilities of a woody cutting (e.g., the presence of resin ducts or secretory cavities, whose differentiating margins may be the site of meristemoid formation).

B. Physiological basis of rooting of cuttings

The process of rooting is governed by various genetical and environmental factors; however, the physiological changes occurring in plants during rooting can not be over sighted. Several physiological processes occur during the rooting of cuttings. These are as follows:

a) Growth regulators: Various classes of growth regulators, such as, auxins, cytokinins, gibberellins, abscisic acid and ethylene influence rooting of cuttings, which have been discussed as under;

Hormone: The first naturally occurring auxin, indole-3-acetic acid (IAA) was identified in 1934 to induce root formation in cuttings. Later, it was found that indole butyric acid (IBA) and naphthalene acetic acid (NAA) are more effective in inducing rooting in cuttings. Cytokinin group of hormones is responsible for growth and differentiation of cells. Cuttings having high cytokinin levels are difficult-to-root. Similarly, if cytokinins are applied artificially, root initiation process is inhibited. Gibberellins and ethylene at low concentration promote root initiation and *vice versa*.

Vitamins: Endogenous optimum level of vitamins or exogenous application of vitamins gives beneficial effects in rooting initiation in cuttings. thiamine, pyridoxine, niacin and biotin vitamins, if applied with some auxin (indoleacetic acid), stimulate root initiation.

Endogenous rooting inhibitors: Cuttings of some plants may not root mainly because of presence of natural occurring inhibitors. For example, the cuttings of Vitis *berlandieri* do not root easily as compared to *V. vinifera,* due to presence of higher amount of inhibitors. The leaching of inhibitors by placing the cuttings in water, before planting, enhances the root initiation and subsequent growth and development.

b. Presence of buds and leaves: The presence of leaves on the cutting exerts a strong stimulating influence on the root initiation process in cuttings. The leaves produce carbohydrates, besides being powerful auxin producers, which are undoubtedly translocated downward to the base of stem, where these promote root formation.

c. **Rooting co-factors:** Rooting co-factors are natural occurring substances that appear to act synergistically with auxins, primarily indoleacetic acid (IAA), for root formation in cuttings. Bouillenne and Went were the first to name the root-forming factor of leaf, cotyledon and buds as rhizocaline in 1933. Later, a new hypothesis was proposed by Bouillenne and Bouillenne-Walrand in 1955, highlighting that rhizocaline is a complex of three components, *viz.*, i) A specific rooting factor, which is translocated from buds, leaves, characterized chemically as ortho-dihydroxy phenol, ii) A non-specific factor (auxin) in low concentration, which is also translocated from leaves and the buds, and iii) A specific enzyme, probably polyphenol oxidase, present in pericycle, phloem or cambium. According to this hypothetically scheme, ortho-hydroxyphenol reacts with auxin, giving rise to rhizocaline. The formation of rhizocaline leads to the reactions involved in the root initiation process (**Figure 12**). Another hypothesis was proposed by Hess (1968) indicating that a number of co-factors interact with IAA, resulting in IAA co-factor complex, which subsequently initiates rooting process. Usually, the easily rooted plants have a large number of co-factors and in higher concentration than in difficult-to-root plants. One of these co-factors (No 4), consists of highly active substance, characterized as oxygenated terpenoids. Another factor, No 3 was identified as chlorogenic acid, Later, it was postulated that phenolic compounds, especially catechol, reacts synergistically with indole acetic acid in root production. This oxidation of ortho-dihydroxy phenol is one of the first steps leading to root initiation process. Similarly, abscisic acid was also found to be associated as one of the co-factors in root initiation as it promotes root initiation. The action of phenolic compounds in root initiation is mainly in protecting the natural occurring auxin-indoleacetic acid from destruction by indole acid oxidase. Rooting co-factors have been reported to occur in many plant species like *Hedera helix*, Crab C, MM_{106} and M_{26} apple rootstocks. These were also found in easy to root Old Hume pear cultivar but not in difficult-to-root Bartlett.

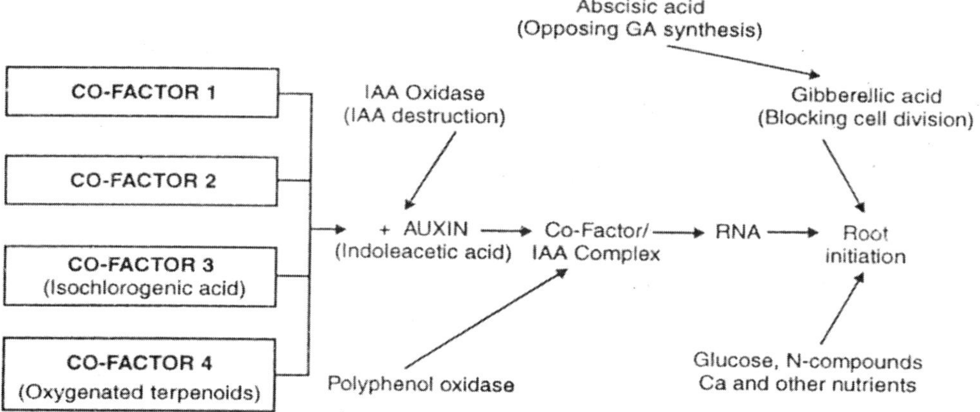

Fig. 12. Schematic representation of involvement of rooting co-factors in rooting process of cutting

d. Nutritional factors: The nutrient status of plant from which cuttings are taken, also plays vital role in root initiation process of the cuttings. However, among different, nutrients, nitrogen and carbohydrate reserve is important. Low nitrogen and high carbohydrate reserve usually favour root initiation process compared to high nitrogen and low carbohydrate reserve. High N content usually favours luxuriant growth but hinders root initiation process, and extreme N deficiency also hinders the rooting process. Thus, high C:N ratio should be maintained in the stock plant before taking cuttings from them.

Phases of rooting

Four phases of rooting have been recognized,. which are discussed as under;

1a) **Physiological Inception:** During this phase, no visual change is noticed in cuttings. This entails internal physiological changes that are a prerequisite for the subsequent developmental changes. Probably, this phase involves hormonal changes, gene deregulation, etc. This is likely to be the stage of cuttings, where hormonal treatments (rooting powders) have their effect.

b) **De-differentiation:** At this stage, cytoplasm becomes dense and nucleus enlarges. Further partially differentiated cells, usually parenchyma, become less specialized and de-differentiate to a meristematic state, which is capable of cell division. Some researchers consider that at this stage, cells become competent to respond to the rhizogenic signal of auxin. Formation of starch grains can also be seen during this phase. Growth regulator like gibberellin is inhibitory to the process of de-differentiation.

2) **Induction or Initiation:** The initiation last from the first cell division to formation of a primordia. By the action of auxin, some cells become determined to root formation at this stage. This begins with the first cell division, the formation of a group of unorganized and non-differentiated cells, followed by the development in to root primordia. The starch grain can be seen degraded, when observed under a microscope. Hormones like cytokinins and ethylene are inhibitory to induction of rooting.

3) **Differentiation:** During differentiation, emerging primordia develops apical zonation and vascular differentiates. The emerging primordia take on the morphological characteristics of a root. Application of auxin at this stage has been found inhibitory.

4) **Elongation:** The emerging root elongates through the parent plant body towards the exterior.

5) **Maturation:** The root emerges and continues growth and development which is a characteristic of a typical root for that particular plant. For all intents and purposes, it behaves like a primary root apex.

Factors affecting rooting of cuttings

The ability of the rooting of cutting in different plant species varies widely. The rooting of cuttings is influenced by both internal and external factors. These factors are briefly described as under:

1. **Physiological condition of the mother plant:** It can exert a significant influence on rooting of the cuttings. In many plant species, the cuttings taken in the early morning hours, when plants are in turgid condition, usually root better than the cuttings taken from water deficient plants. Cuttings taken from nutrient deficient plants rarely root better. Restriction of downward movement of carbohydrates, auxins and other possible root promoting factors, also result in better rooting of cuttings. Thus, the girdling and etiolation of shoots prior to their use as cuttings is useful for stimulation of rooting.
2. **Juvenility factor:** The ability of cuttings to form adventitious roots decreases with the increase in the age of the plant due to presence of more rooting inhibitors and decreased level of polyphenols in the older plants. In fruit plants like mango, apple, citrus, pear, peach, etc., the cuttings often root better if taken in juvenile period (pre-bearing period) and the cuttings taken from old, non-bearing plants rarely root better. The juvenility is induced artificially in mature old plants by heading back, spraying GA_3, stooling, disbudding, etc.
3. **Type of the wood:** Cuttings taken from lateral shoots root better than those taken from terminal shoots because lateral shoots have more stored food, facilitating better rooting in the cuttings. In hardwood cutting, rooting is more in basal portion of the shoot than on the middle or top portion because accumulation of food and root promoting substances are more in the basal parts. Further, cuttings taken from a vegetative shoot root better than those taken from flowering shoot. Similarly, cuttings are made with a heel (a small piece of old wood) or mallet (a short segment), usually initiate rooting faster due to presence of preformed root initials in them.
4. **Presence of leaves and buds:** The presence of leaves and buds is essential for the rooting of cuttings as they are the primary source of carbohydrate and auxin synthesis in plants which are transported to the basal portion of the cutting for root initiation. Usually, growing buds promote rooting and dormant buds inhibit it. However, in some cases, the leaves are removed to some extent to reduce the loss of water due to transpiration.
5. **Presence of viruses:** The process of root initiation in cuttings taken from plants infected with viruses is inhibited or reduced as compared to those taken from virus free stock plants. It is particularly true with apple and Chinese gooseberry, where virus free clones have been reported to root better than the infected stocks. Viruses

not only reduce rooting percentage but also the root number.

6. **Position of the basal cut in the cutting:** Sometimes, the position of the basal cut with reference to the node has been found to affect root initiation in the cuttings. In some plants, rooting is better when the cut is made just above or below the node, in others rooting is better if cut is made at the node, and still in others, position of the cut has no effect on root initiation process.

7. **Time of the year:** In some instances, season of the year, when cutting are produced, has significant influence on rooting of the cuttings In deciduous plants, the hard wood cuttings could be made in dormant season and semi-hard wood or soft wood cuttings in the growing season. The evergreen plants usually have one or more flushes of growth in a year and thus cuttings should be prepared at various times in relation to growth flush but more especially in spring or later fall, depending upon the species. However, the response of the rooting of cuttings in different plant species may vary widely.

8. **Treatment of cuttings:** Various treatments to the cuttings before planting have been demonstrated to be useful for root initiation. These are:

   a. **Use of growth regulators:** Among growth regulators, indole butyric acetic acid (IBA) has been found to be the most ideal compound for promoting rooting in cuttings in most of the plant species. The concentration varies from plant-to-plant and type of the cuttings used. Naphthalene acetic acid (NAA) and 2,4-D (2,4-dichlorophenoxy acetic acid) are other root promoting auxins with varying degree of success. Plant growth regulators may loose their strength as these may be destroyed by bacteria (*Acetobacter* destroys IAA but not IBA, NAA or 2,4-D) or light. Further, IAA is highly sensitive to light and is destroyed by strong sunlight though NAA and 2,4-D are stable. In propagation with leaf or root cuttings, production of buds and shoots is necessary. For this, treatment of cuttings with cytokinin, such as, kinetin or 6-benzyl adenine (BA) is useful.

   b. **Mineral nutrients:** Different mineral nutrients may also have significant influence on rooting process of cuttings of some plant species. Treatment of cuttings with nitrogenous compounds (organic and inorganic) usually promotes rooting in cuttings in several plants. Among different organic forms of nitrogen, aspergine and adenine are most effective in doing so. Boron also plays important role in rooting process of certain plant species. It promotes root growth rather than root initiation. The combination of nutrients (N, B) with auxins (IBA) is the most effective treatment for root initiation and development process in many plant species. Concentrations, of course, vary from plant-to-plant and type of the cutting to be used.

c. **Fungicides:** In some instances, the rooting of cuttings initiate at a better rate but their survival is low because these are attacked by pathogens. Thus, treatment of the cuttings before planting, with fungicides like captan and benomyl, gives better results. The fungicides may be used as a powder dip, following IBA treatment or these may be mixed with IBA powder before treatment. Out of these two fungicides, captan is more suitable for treatment of cuttings because it does not decompose easily and has long residual effect.

d. **Wounding:** Basal wounding of the cuttings, in some instances, is helpful for inducing rooting in cuttings. Wounding is helpful in cuttings having older wood at the base. Wounding promotes rooting in cuttings in several ways: i). It tears sclernchyma rings of tough cells in the cortex, exterior to the point of origin, ii). It helps in better absorption of moisture, growth substances from the rooting medium, iii). Due to higher excretion of hormones in the wounded area, respiration rate is increased, iv). Division of cells in the wounded area and cells adjacent cells is faster, and v). Wounded cells release ethylene, which act as root promoting hormone. All these conditions induced by wounding, which helps in better root initiation process.

e. **Etiolation treatment:** Etiolation is the oldest horticultural practice in which light is excluded from the plant or its part for sometime during the growing season. Sachs, a renowned scientist was the first to demonstrate the effect of light exclusion on adventitious root initiation during 1864, which was later, confirmed by many other research workers. It is thought that exclusion of light reduces the photo-decomposition of naturally occurring auxins and resulting in accumulation of auxins, which ultimately synergies the root initiation process. In the actual practice of etiolation, the basal portion of the newly emerged shoots is kept under complete darkness by covering them with a black polyethylene and the terminal portion is allowed to grow in normal light conditions. After sufficient time, which varies from species to species, the shoot is cut down, its basal portion is treated with auxins (IBA) and cuttings so prepared are planted in the nursery. The trench and mound layering and stooling methods of propagation in which basal portion of the shoots are kept in darkness by soil cover, are based on the principle of etiolation. Experiments have shown that etiolation is useful in inducing better rooting in cuttings of difficult-to-root plant species like mango, jackfruit and avocado.

f. **Girdling:** In girdling, a ring of bark, preferably 2.5 to 3.0 cm is removed from the base of the shoot, which is to be used for the preparation of cuttings.

It can also be done by making a notch or tying a wire around the shoot. This process helps to initiate the rooting process much early in some plant species. Girdling basically blocks the downward movement of carbohydrates, hormones and other root promoting factors. When to girdle and from where to girdle, differs widely among plant species. However, cuttings of mango, litchi, guava, citrus root better if the shoots are girdled about 10-15 days before preparing the cuttings from them.

g. **Association with mycorrhiza:** The addition of mycorrhizal fungi into rooting substrate can increase rooting im the cuttings. This has been observed in fruit crops like blue berry, huckle berry, grape, pomegranate etc.; besides, several ornamentals. Therefore, addition of AM fungal inoculum to rooting substrate could be an effective strategy for the nursery production of mycorrhizal plants.

9. **Environmental conditions:** Environmental conditions during rooting, like water relations, light, temperature and rooting medium, may also affect the rootings in cuttings. For better rooting in the leaf cuttings, the turgidity of the leaves has to be maintained. If rooting medium is water-deficient and the leaves continue transpiration, it results in loss of turgidity. A day temperature of 21-24^{0}C and a night temperature of 13-15^{0}C are considered quite satisfactory for rooting of cuttings. Excessively high or low temperature inhibits root initiation in cuttings. Light is the major source of energy in photosynthesis. In rooting of cuttings, the formation of carbohydrates is most important. Light intensity, day length or light quality affect rooting of cuttings. Usually, low light intensity gives better rooting in cutting than obtained from stock plants grown at high light intensity, thus etiolated shoots root better than non-etiolated shoots. The orange red light of the spectrum seems to favour rooting of cuttings than the blue region. Further, red light (680 nm) has more inhibitory influence on rooting than the blue or far red light spectrum. Some plants manufacture good carbohydrate under short day conditions and others at long day or neutral day conditions and thus rooting in the cuttings is influenced.

10. **Rooting medium:** An ideal rooting medium must provide sufficient porosity to allow free passage of air, should have better water holding capacity, yet well drained and must be free from harmful pathogens. It has to perform three important functions *viz.*, 1. It has to hold the cutting properly, 2. It has to provide adequate moisture to the cuttings and 3. It has to permit free air passage to the base of the cuttings. Thus, type of the medium may also significantly influence rooting of cuttings. Some cuttings when rooted in sand produce long, unbranched and brittle roots but produce well branched, slender and flexible roots in perlite or peat mixtures. In addition to the texture and structure, the pH of the rooting medium above and

below 7 may be harmful for root initiation process in cuttings of the most of plant species. Generally a pH near 7 is considered ideal for rooting process in the cuttings. Further, the level of exchangeable calcium in the rooting medium, especially peat moss should not be high as it may also affect rooting process adversely. Similarly, the availability of proper concentration of oxygen (1-10ppm) in the medium is essential for root initiation process.

Cuttings and their types

Any vegetative part of the plant detached from mother stock and treated for induction of root and shoot is considered as cutting. Cuttings are prepared from vegetative portion of plants such as stem, leaf and root. On the basis of plant part used and relative position on a plant, cuttings are classified in various groups. Different types of cuttings used by plant propagators are as described below:

A. Stem cuttings:

Stem cuttings is considered the most convenient and popular method of plant propagation, next only to seed propagation. Most cuttings fall into this category. A stem cutting is any cutting taken from the main shoot of a plant or any side shoot growing from the same plant or stem. The facility with which different species can be propagated in this way is largely determined by experience. There are, however, a few general considerations, which help in selection of suitable cuttings. Cuttings are detached from the mother plants and the source of food supply is excluded. Therefore, it is essential to have sufficient reserve food to keep tissues alive until roots and shoots are produced. The shoots with high carbohydrate content usually root better. To maintain high carbohydrate content in a shoot, ringing or notching stems down to the wood are useful practices. As a general rule, cuttings from young plants root better, but, if older shoots of the plants are cut back hard, very often they can be induced to produce suitable shoots for rooting. Stem cuttings may be divided into four groups based on the age or maturity of the tissue when the cutting is taken. These are hardwood, softwood, semi-hardwood and herbaceous cuttings as follows;

i) **Hardwood cuttings:** Cuttings made from past season's growth or wood that has matured and lignified is known as hardwood cuttings. It is the simple and cheapest method of plant propagation. Weak, fast growing shoots with long internodes should not be used for this purpose. Length of cutting varies from 10 to 45 cm and it should have at least two buds. While preparing the cutting, a straight cut is given on the base of shoot below the node while a slanting cut, 1 to 2 cm above the bud at the top of the cutting. This helps in maintaining the polarity of the shoot while planting and if rain occurs, water does not accumulate on the tip of the cutting, which saves the cutting from fungal infection. Fruit plants commercially propagated through hardwood cuttings are grape, fig, sweet lime, pomegranate, mulberry, plum, olive, apple, etc. In some cuttings (hammer cutting), a piece of twig is cut together with the stem because some plant cuttings, like gooseberry cuttings, are difficult-to-root and the additional piece of twig helps to develop root system.

ii) **Semi-hardwood cuttings:** The cuttings taken from partially matured, slightly woody shoots are called as semi-hard wood cuttings. Such cuttings are usually taken from growing terminal shoots just after a flush of growth. This type of cuttings is mostly used in evergreen fruit plants like mango, guava, lemon, jackfruit, olive, etc. Length of cutting varies from 7 to 20 cm and two-to-four leaves should be retained on the top of cutting. Treating the cuttings with 5,000 ppm IBA before planting gives better results.

iii) **Softwood cuttings:** Cuttings made from soft, succulent and non-lignified shoots of woody perennials are called as softwood cuttings. The size of cutting varies from 10-15 cm and few leaves to be retained on cutting. Before planting, treatment with auxins (IBA, NAA) is beneficial. The best time for preparing softwood cuttings is late summer. Usually terminal portions are taken, but soft basal and middle position can also be used (**Figure 13**). These cuttings are root easier and quicker than other cuttings. Softwood cutting are used for the propagation of apple, peach, pear, plum, apricot and cherry plants under mist unit. Likewise, Juniper, azalea and lilac are also propagated easily by softwood cuttings.

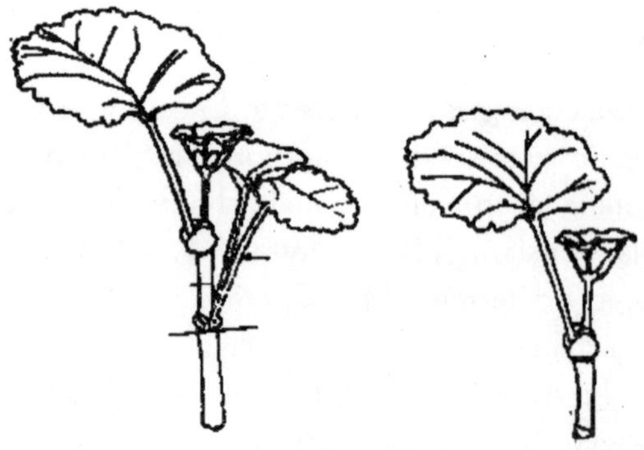

Fig. 13. Preparation of softwood cuttings from plants.

iv) **Herbaceous cuttings:** The terminal leafy portion of the stem of a plant is used for preparing herbaceous stem cuttings. These cuttings are soft and succulent with length varying from 7-15 cm. These cuttings are rooted under the same conditions as the softwood cuttings, requiring high relative humidity. Bottom heat is also useful for initiation of rooting process. Under proper conditions, the rooting is very high. However, these are liable to wilt soon if proper humidity is not maintained. Therefore, much attention is required for working with soft herbaceous cuttings. Herbaceous cuttings of some plants exude a sticky sap (as in geranium, pineapple, cactus etc.) that interferes with root initiation process. In such cases, basal ends of cutting should be allowed to dry for few hours before planting. Generally, fruit plants are not propagated by herbaceous cutting. However, some flowering plants like chrysanthemum, dahlia, coleus, carnation etc., are propagated through herbaceous cuttings.

B. Root cuttings

Propagation by means of root cuttings is also a simple and cheap method of vegetative propagation in species, which are difficult-to-propagate by other methods (**Figure 14**). In general, the plants, which produce suckers freely, are easily propagated by root cuttings. It is found that the plants, which are difficult-to-propagate by stem cutting are sometimes easy to be propagated by root cuttings. For preparation of root cuttings, roots, which are of 1cm thickness and 10-15cm long are cut in to pieces. The best time for taking root cuttings is late winter or early spring, when the roots are well supplied with stored food material but before the new growth starts. However, in temperate fruits, root cuttings are prepared in the month of December and are kept in warm place in moss grass or wet sand for callusing and are then transplanted during February-March in the open beds. While using root cuttings as the propagating material, it is important to maintain the correct polarity. Thus, to avoid their planting upside down, the proximal end of the cutting should have straight cut and distal end a slanting cut. The proximal end should always be kept up, while planting, insert the cuttings vertically so that the top is above the soil level. However, in some cases, horizontal planting gives satisfactory results as in sweet potato. Blackberry and raspberry are commercially propagated by this method. However, kiwifruit, breadfruit, fig, rose, mulberry, apple, pear, peach, cherry and persimmon are also propagated by root cuttings. In some variegated foliage plants like pelargonium, bougainvillea, the new plants more often loose their variegated forms when propagated through root cuttings.

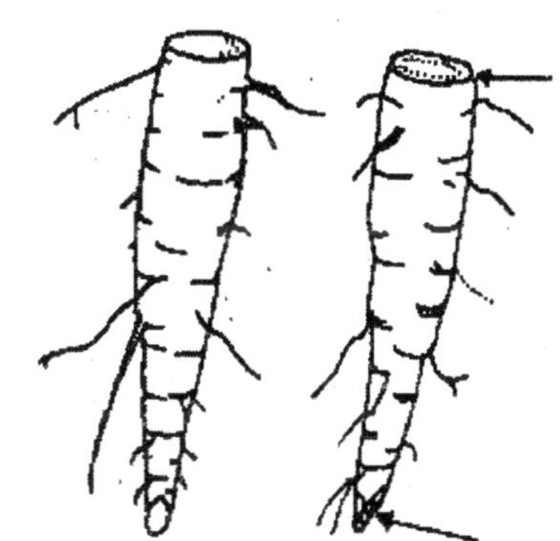

Fig. 14. Propagation by root cuutings.

C. Leaf cuttings

In some plant species, the thick and fleshy leaves have the capacity to produce new plants. In these species, leaf blade or leaf sections are used for further multiplication. The roots develop on the leaf blade or leaf section. Some greenhouse plants can be propagated by means of various types of leaf cuttings. In leaf cuttings, the formation of root and shoot is essential to get a complete plant. In many cases, only roots develop but not the shoot system (camellia, rhododendron etc.), as a result of which, the cuttings die. Leaf cuttings should be made from fully expanded recently mature leaves and old leaves may be avoided. Leafy cuttings require environmental conditions similar to herbaceous cuttings but for high success with leaf cuttings, relative humidity should be a little high. In some plant species, particularly variegated *sansevieria*, the new plants don't come true-to-type if propagated by leaf cuttings. The different types of leaf cuttings used for propagation are leaf petiole cuttings, leaf sections and leaf slash etc. In leaf petiole cutting, complete leaf with its stalk is used for propagation. This is an easy and simple method of plant propagation but plants produced by this method are very few. Saintpaulia and kalanchoe are propagated through leaf petiole cuttings.In leaf sections, a section of the leaves about 1-10 cm long is cut at right angle to the veins and planted in the rooting medium by maintaining the polarity. *Sansevieria* is mainly propagated through leaf sections.Leaf slashing is commonly used in plants having capacity to regenerate new plants from the leaf veins. In this method, the cut leaf is placed flat on the medium and the leaf is pinned down with a thin wire. The new root and shoot system develops on each cut vein. It is commonly followed in *Begonia rex.*

D. Leaf bud cuttings

Propagation through leaf bud cuttings is partially useful in species where leaves develop root system but die because of non-development of shoot system. Leaf bud cuttings are particularly useful when planting material is scarce because the each node in leaf can be used as cutting. Leaf bud cutting should preferably be prepared during growing season because buds, if enters in dormancy, may be difficult to force to active stage. A leaf bud cutting consists of a leaf blade, petiole and shoot piece of stem with attached axillary bud of actively growing leaves. In leaf bud cuttings, 1-15 cm stem portion is used when propagating material is small. These cuttings should be made in growing season. It is useful method of propagation in black berry, raspberry, lemon, camellia etc.

In literature, one would find different names for cutting, primarily used to designate from the part or portion of plant these are taken. The different kinds of cuttings are also used by the propagators and nurserymen, which are described as under:

1. **Basal cuttings:** A basal cutting is a young shoot cut or gently pulled or removed from the base of a parent plant. Generally, such cuttings need little preparation and only the removal of a fewer lower leaves is required before planting. They should be taken from young plants. A cutting length of 4-8 cm is most ideal. These require some form of protection from drying out while rooting. Many perennials and alpine trees are propagated from basal cuttings.
2. **Heel cuttings:** A heel cutting is usually a side shoot of a hardwood or semi-hard wood cutting pulled from the parent plant complete with a heel of bark from the main stem. *Jasminum* roots better from such cuttings. These should be prepared and planted as ordinary hardwood or semi-hardwood cuttings without doing any-

thing to the heel at the base of the cuttings, other than trimming the loose ends with a sharp knife before planting.

3. **Bud cuttings:** Taking a bud cutting is similar to preparing a bud for budding, the main difference being that with the cutting, the leaf is retained to the slice of the stem. A semi-hard wood cutting should be choosen to supply the bud from the current season's growth in the late summer. A slice of stem around 2.5 cm in length is cut from the parent plant, each cutting containing a leaf and a leaf axil containing a bud. The cutting is planted in such a way that the bud and leaf are just above the surface of soil. Rose and camellias are among the plants propagated through bud cuttings.

4. **Eye cuttings:** This is very successful method of propagation of vines. An eye cutting is a hardwood stem cutting taken from the parent plant during the dormant period of autumn or winter. The best cutting is one, which is 4-5 cm having an eye or a bud. The pruned wood of vines is cut into many sections, each containing a bud. With the sections so cut, a strip of bark is removed from the opposite side of the stem to the bud and the cutting is pegged down in the soil. With the cutting lightly pegged down, sprinkle water and press the cutting so that only the bud protrudes above the compost/sand. A bottom heat of 24 °C is required for rooting. With this temperature, the compost may dry out rapidly, so check periodically to make sure that it remains moist for which water may be given frequently.

5. **Internodal and nodal cuttings:** The node is that part of plant where the leaf stalk or petiole joins with the stem. Most cuttings root better if they cut across in a straight line immediately below a node, and are therefore termed as nodal cuttings. Nodal cuttings are, specially, advantageous to those plants having hollow stems, as the node generally leads to a branch of solid tissue throughout such stems. Some plants though appear to even root better if the stem is cut midway between a pair of nodes. These cuttings are therefore termed as internodal cuttings. Clematis is one plant propagated through internodal cuttings.

6. **Irishman's cuttings:** An Irishman's cutting is a basal cutting, which is pulled from the crown of a parent plant having a few roots already attached to it. It is perhaps a mild form of division. This type of cutting is quite often used to propagate early flowering chrysanthemums and some herbaceous perennials.

7. **Piping cuttings:** A piping cutting is simply a stem cutting taken from the tip of carnation. It is very easy to take the tip of any non-flowering shoot being suitable. For taking piping cutting, hold the selected stem with one hand and pull the tip carefully with the other. This piping is just dipped into a rooting compound and dibbled into sandy compost for rooting. A cold frame is suitable for proper rooting in these cuttings. Different types of cuttings and their relative position on a plant is depicted in **Figure 15**.

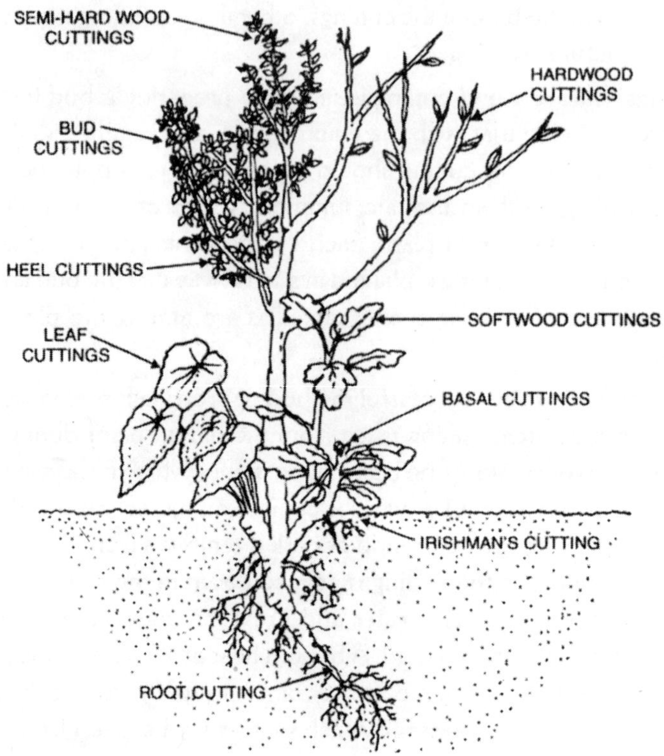

Fig. 15. Different types of cuttings and their relative position on a plant.

6 Propagation by Grafting: Principles and Practices

Grafting of plants is an ancient technique, which was supposed to be known to the Chinese as early as 1,000 B.C. Centuries later, it was employed widely by the Romans. First-century Romans used cleft grafts by placing a shoot in a split in the wood. While working with roses and fruit trees, the Romans discovered that rootstock influences the vigor and size of a plant. Grafts and buds were tied with bendable twigs or bark, including willow bark. Leaf fibers of the raffia palm tree later replaced willow bark. The Romans used to seal the grafts with chalk mixed with cattle dung or clay. This was later replaced by bees wax, still used by some gardeners. The technique of grafting was discussed in detail by historic writers including Aristotle and Theophrastus and it became very popular in Europe during the Renaissance (1350 to 1600 AD). By the 19th century, more than 100 grafting techniques were described in literature, including many that are widespread today. In his 1821, the book "Monographic des Greffes," the French botanist Andre Thoulin described 1,119 kinds of grafting. In history, grafting has led to development of new industries from scattered wild individual plants. An interesting example is the commercial cultivation of pecan tree, which served as an important food source for Native Americans. In the mid-1880s, efficient pecan grafting techniques transformed a chance wild harvest into a commercially important crop. Today, grafting has become an inevitable part of fruit crop propagation industry, as this is, by-and-large, the most suitable technique to maintain characteristics of a clone.

Grafting is an asexual propagation method in which parts of two different plants are joined so that they continue their growth as one plant. To accomplish this, one of the two parts serves as the bottom part, which is in contact with the soil and is called the 'rootstock', and the other top part or scion. In this union, the plant material used as the scion is being propagated and is usually the only one allowed to grow. In effect, the scion becomes the new shoot of the plant union and stock serves as the root, conducting nutrients and water across the graft junction and into the shoot. Many horticultural plants are propagated by grafting. In grafting, the desired cultivar can be raised on other plants for achieving the desirable benefits. Propagation by grafting also involves some principles, which are important for use to learn. In order to realize the complete benefits of grafting, it is inevitable to ensure the good contact of the cambium or growing layer of the scion and rootstock. The cambium is located just below the bark and is a layer of actively dividing cells, which are responsible for of the conducting vascular system of the stem, wood and bark. The cambium plays a pivotal role in the union between the scion and rootstock. Therefore, the success of grafts rely upon the fact that how well the cambium of the scion and rootstocks are aligned for a good union to occur.

It is necessary to select healthy and vigorous rootstocks and scions, apparently free from any disease infection or insect infestation. The scion material should be taken from a new flush with buds, which are swollen but have not yet opened. To encourage bud development on scion, *precuring* is carried out by removal of leaves, while the scion is still attached to the mother plant. *Precuring* is usually, done, 15-20 days before severing of scion from mother tree. Removal of leaves of the scion aids in reducing water loss after grafting. Care should also be taken to adequately water the rootstock before grafting. Immediately after grafting, the plants should be kept in a shaded environment with relatively higher humidity.

Grafting has certain advantages and disadvantages over other methods of propagation. These are as under:

Advantages

Grafting has following advantages over the other methods of propagation:

- It is a better method of propagation for plants, which are not easily propagated through cuttings and layering.
- Grafted plants are precocious in bearing i.e., they come in bearing much early as compared to seedling plants. Therefore, this attribute of grafting can be exploited for shortening of breeding cycle.
- The desirable benefits of a rootstock are best utilized through grafting.
- Graft incompatibility can be overcome by using an interstock, which is compatible with both rootstock and scion.
- Old or declining orchards of inferior variety can be rejuvenated with superior cultivars by top working.
- A number of varieties can be grafted on one plant by grafting.
- The damaged plant parts can be repaired easily by grafting.
- Grafting is helpful in pollination of some dioecious plants and overcoming self-incompatibility in temperate fruit plants like apple, pear etc. The cross-pollination can be ensured by grafting a scion from a male plant/pollinizer onto the female/self incompatible plant.
- Viral diseases are studied through indexing of plants by grafting.

- Special forms of plants, such as, 'weeping' or cascading' can be obtained by grafting. For example, Special forms of plant growth such as "weeping" cherries etc., can only be obtained by grafting onto Mazzard cherry rootstock (*Prunus avium L.)*.
- Perpetuation of clone can be achieved by grafting for example, clones of numerous species of conifers cannot be economically reproduced from vegetative cuttings because the percentage of cuttings that root successfully is low. Colorado blue spruce (*Picea pungens*), Koster blue spruce (*Picea pungens* var. Kosteriana), and Moerheim spruce (*Picea pungens* var. Moerheimii) are commonly grafted onto Norway spruce (*Picea abies*, Karst.) or Sitka spruce (*Picea sitchensis*) rootstock to perpetuate desirable clones.
- Grafting has been widely used in physiological and genetic studies to determine the transfer of mobile elements in plants. This includes translocation of alkaloids and secondary metabolites; transfer of the flowering stimulus (florigen); transfer of a potato tuberization stimulus (tuberin); transfer of growth substances such as cytokinin from roots-to-shoots; and the transfer of RNAs.
- A unique use of grafting has been harnessed in Taiwan pear industry. The Japanese pear, or nashi (*Pyrus pyrifolia*), is a favorite pear in Taiwan, but cannot be grown in the subtropical areas in want of sufficient chilling. This is overcome by flower bud grafting. After trees in Japan have received chilling nearly sufficient to break dormancy, budwood is harvested and shipped to Taiwan, where it is held at 2° to 4°C to complete the chilling requirement. The pre-chilled scions are grafted into 'Heng Shan' pear trees and produce two to four marketable fruits per cluster. The technique benefits Japanese growers, who obtain an additional source of revenue from exporting buds, and Taiwanese growers, who profit by early crops of marketable fruit that receive very high prices.
- Mukibat grafting of cassava, developed in 1958, is an example of horticultural innovation by a peasant subsistence farmer Mukibat, who lived on the island of Java, Indonesia. He grafted an inedible relative of cassava, the Ceara rubber tree (*Manihot glaziovii*), onto a cassava (*Manihot esculenta*) stock. The resulting yield of cassava tubers was dramatically increased to 30% to 100% or more, compared to ungrafted cassava.

Disadvantages

Grafting has certain limitations, as described briefl hereunder:

- All the plant combinations are not compatible and thus it is not possible to follow grafting in such cases.

- Grafting techniques are successful only in dicots and gymnosperms. Monocots do not respond to grafting, primarily because of absence of cambium layer.

The success of grafting primarily depends on the successful union of stock and scion. However, there is great variability in different plant species in their ability for forming a successful graft union. Some plants can be grafted easily, while the others with great difficulty. There are six general requirements for any successful grafting operation. These are:

(i) Compatibility: The two plants to be grafted must be compatible. Even though this union is physical, the two plants should be as closely related genetically as possible for success. In the Table 3, the limits of grafting has been discussed;

Table 3: Grafting limits in fruit crops.

S. No.	Graft combination	Remarks
1.	Within a variety	Good
2.	Between varieties	Good
3.	Intrageneric (different species but under the same genera)	Good. For example Japanese plum cultivars can successfully be grafted on European plum cultivars but the reciprocal grafts are not successful.
4.	Intergeneric (between different genera)	Not very frequent. Pears can be grafted on to quince. Oranges on *Poncirus*.
5.	Between families	Not reported in perennial fruit crops; however, successful union was found between white clover (Leguminosae) as scion and sunflower (Compositae) as stock.

(ii) Matching of tissues: The cambium layer of the two matched plant parts, the scion and the rootstock, must come in close contact and be held tightly together (**Figure 16**). The cambium must not be allowed to dry out.

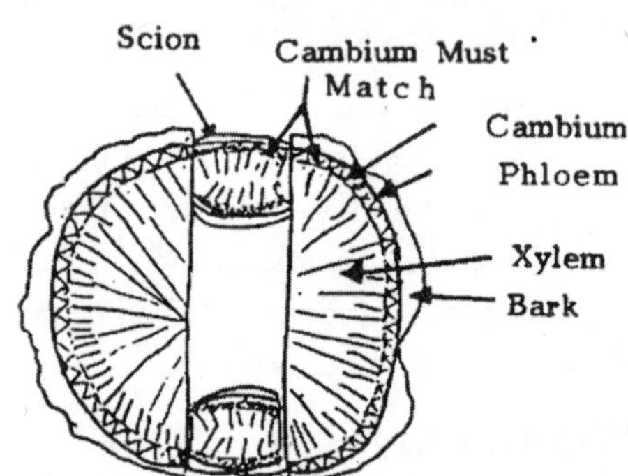

Fig. 16. Matching up of scion and stock cambium for successful union.

(iii) Physiological stage: For a successful graft, both the scion as well as stock must be at the proper physiological stage. For instance, they should not be in stress and should not be over aged.

(iv) Water proofing: Immediately after the graft is made all cut surface must be covered with a water proofing material such as grafting wax, plastic ties, rubber ties etc.

(v) **Season:** The grafting operation should be done at the right time of the season when both stock and scion are at proper stage of their maturity and growth. Generally, it is recommended that time when cambium cell are in active division state, is considered as best for grafting. This can be judged by observing the slippage of bark. The growth stage, thickness, and physical state of scion influence widely the success of grafting.

(vi) **Other factors:** The grafts should be protected from the vagaries of nature and attack of insect-pests and diseases. Similarly, off shoots appearing below the graft union should be removed or rubbed off as they appear.

Plants suitable for grafting

Broad leaved dicotyledonous and conifers are suitable candidates of grafting. Monocotyledonous plants such as bamboo, grasses or lilies can not be grafted. The reason being the absence of cambium in such plants, which play a pivotal role in the union of different parts (scion and stock). Unlike dicots, which have continuous cambium layer just beneath the bark, monocots have vascular bundles, which make the matching of two plant parts impossible. Moreover, there are always better chances of formation of a successful union if both the partners (stock and scion) have the better capacity of producing parenchymatous cells near the cambial layers.

Graft union

Grafting involves the union of a root system (called the rootstock/understock) with a shoot system (the scion) in such a way that their cambiums are aligned and subsequently grow to develop one composite plant. The successful graft union is the basis of grafting. The formation of graft union is basically a process of healing of wounds. A graft is successful if healing is complete and vascular transport is restored between the two parts (scion and rootstock). In this process, the parts of graft that are originally prepared and placed in close contact don't themselves move about or grow together. The union is accomplished entirely by cells that grow and develop after the actual grafting operation has been done. Further, in a graft union, there is no intermingling of the cell contents of both stock and scion and cells maintain their own identity.

Formation of graft union

The main steps involved in the formation of graft union include, i) wounding response and the formation of callus tissue consisting of parenchymatous cells by the exposed cells of cambium of both the partners i.e. stock and scion, ii) inter-locking of cells and callus cells of stock and scion, iii) differentiation of certain parenchymatous cells into new cambium cells, and iv) formation of new vascular tissues by the new cambium for establishing a vascular connection for the translocation of water, nutrients and other metabolites.

Steps in graft union formation

1. Wound response and callus formation

In response to wounding (result of cuts made on scion and stock by knife), a necrotic layer forms both on scion and stock, which is followed by the production of undifferentiated callus tissue from uninjured parenchyma cells below the necrotic layer. Thereafter, callus forms a wound periderm (outer "bark"), which becomes suberized to prevent entry of pathogens. Later, the necrotic layer dissolves.

2. Interlocking of parenchymatous cells

Callus, which mostly comes from scion, proliferates for 1 - 7 days. Callus is mainly comprised of spongy tissue, which penetrates the thin necrotic layer and soon fill up the gap between the stock and the scion and become intimately interlocked. This interlocking of the cells provides a mechanical support to the graftage and also allows the passage of nutrients and water between the stock and the scion. Initially, there is development of a brown line between the callus of stock and scion but it disappears gradually. Similarly, the old vascular tissue of both the components is sealed off with a deposit of gum and thereby accelerates the healing process.

3. Production of new cambial cells

In the callus bridge of the stock and the scion, there are certain parenchymatous cells, which differentiate into new cambial cells after 2-3 weeks of grafting and development of these cells results in the formation of continuous cambial connections between the stock and scion. This is good indication of formation of a successful graftage.

4. Formation of new vascular tissues

After the formation of continuous connection between the stock and the scion through the callus bridge, new cambial activity results in the formation of new xylem towards inside and phloem outside the original vascular cambium of the stock and the scion. The new xylem tissues generally originate from the activities of the scion tissues rather than from the stock. Thus, the production of new xylem and phloem tissues permits vascular connection between the stock and the scion, continuing throughout the life of a grafted plant. It is important that this stage should be completed much before new leaf development on scion otherwise the leaves may wilt and the scion may die. Though, some water can be translocated through callus cells but it is not suffcient enough to support leaves. The cell-to-cell transport takes place *via* plasmodesmata (symplastic transport), which links cells membranes. On the other hand, apoplastic transport occurs between adhering cells.

Factors influencing healing of graft union

The formation of graft union is basically a process of wound healing. This process is influenced by many factors thereby affecting the success or failure of the graftage. These are as under:

1. **Kind of plant:** The ability of two kinds of plants to form a successful graft union is related largely to their natural relationship. Distantly related plants show incompatibility or very slow healing of graft union. The plants like apple, pear, mango, sapota etc., can easily be grafted, whereas, walnut, hickories, oaks are difficult to form successful graft union.
2. **Environmental conditions:** Optimum temperature, moisture and O_2 supply is necessary during and after grafting, for formation and healing of the graft union. Optimum conditions usually vary from plant-to-plant, For example, optimum temperature for grafting of apple is 21-24°C, whereas it is 25-30°C for walnut. For better healing, high humidity is essential to avoid drying of graft union and better growth of the callus. Similarly, optimum supply of O_2 helps in rapid cell division, which is necessary for callus enlargement and proper moisture (irrigation) to the stock plants favours high graft success.
3. **Growth activity of the stock and scion plant:** The condition of the stock and scion during grafting also plays an important role in healing process. If plants are in active growth, they heal soon. However, certain plants (walnut, maple and grape) bleed heavily when cuts are made and such grafts do not heal properly. In mango, defoliation of scion is done before grafting, for encouraging bud break.
4. **Method of grafting:** Grafting technique also influences the healing process of a graft union and it varies in different plant species. A highly successful method of grafting for one plant species may be a complete failure for other species.
5. **Infestation by insect-pests and diseases:** Infection of bud wood /scion sticks by viruses, insect-pests (borer) and diseases may result in poor healing of graft union. Control of these pathogens promotes healing of graft union.
6. **Application of plant bio-regulators and growth promoting microbes:** Plant bio-regulators such as auxin, cytokinins, albumin and ascorbic acid and plant growth promoting rhizobia (PGPR) have been found to promote graftage in various plants like walnut, grape and sweet pepper. Foliar application of bio-regulators/ PGPR promotes/induce the initiation and proliferation of callus and new vascular tissue by promoting cell division and/or cell development at the cut surface of scion stem and improves the survival rate after grafting.
7. **Polarity:** Maintenance of proper polarity is important for proper healing and successful bud or graft union. In budding and grafting, the proximal end of the scion should be inserted into the distal end of the stock (**Figure 17**). However, in root grafting the proximal end of the scion should be inserted in to the proximal end of the root pieces. If the order is reversed i.e., if the scion is inserted upside down, the union may stay alive for some time but don't grow afterwards. Similarly, in nurse-root grafting, the rootstock is knowingly grafted to scion with reverse polarity.

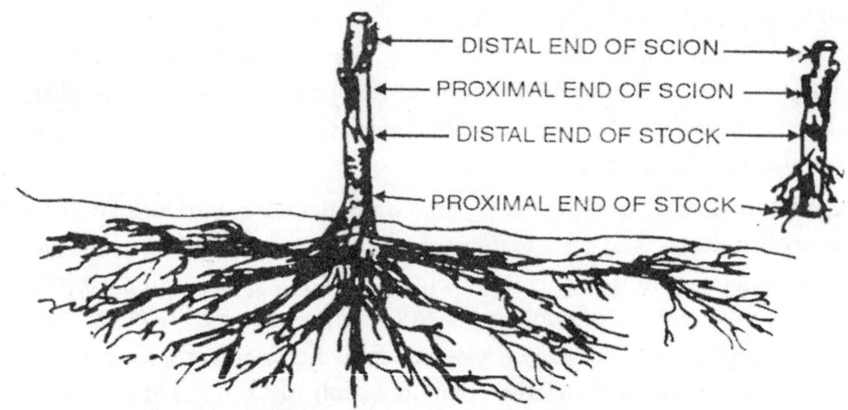

Fig. 17. Polarity in grafting.

Graft incompatibility

The inability or failure of two different plants grafted together to produce a successful graft union is called as graft incompatibility. The ability of two plants to unite successfully into a graft is due to their natural relationship. However, plants differ widely in their ability to produce a successful graft union. For example, some pear cultivars are successfully grafted on quince rootstock, whereas the others may die soon but the reverse i.e., the quince on pear rootstock, always a failure. Similarly, plum grows well on peach rootstock but peach on plum is always a failure. The incompatibility symptoms may appear soon after grafting or may be delayed for several years.

Symptoms of incompatibility

Graft incompatibility symptoms are variable in nature. The following symptoms are known to be associated with graft incompatibility:

- Complete failure to form a successful graft or bud union by some species, varieties or clones.
- Very low percentage of graft success.
- Union takes place and initial growth also occurs but the tree dies pre-maturely either in the nursery or in the field.
- Degeneration of the tissues at graft union, decline in vegetative growth and premature defoliation.
- Marked differences in the growth or vigour of the stock and the scion.
- Appearance of deficiency symptoms or nutritional disorders.
- The plants may have stunted growth.
- There may be yellowing of foliage in later part of the season.

- Appearance of outgrowth at, above and below the graft union (**Figure 18**).
- Excessive swelling at the graft joint, and
- Graft components break off cleanly at the graft union.

However, one or combinations of these symptoms does not mean that the combination is incompatible because the above symptoms may be a resultant of nutritional deficiency, insect-pest and disease incidence or poor budding or grafting technique. Similarly, swelling or outgrowth at, below or above the graft union is not sometimes considered to be associated with incompatibility because some compatible forms also develop such symptoms. For example, Kinnow mandarin budded on Troyer citrange usually develop an outgrowth above the bud union. Similarly, Allahabad Safeda scion of guava also develops such swellings above graft union if grafted on dwarfing 'Pusa Srijjau' rootstock. Thus, the symptoms of outgrowth at or above the graft union are not a reliable indication of incompatibility reaction. However, incompatibility is clearly indicated if the plant breaks off clearly at joint after a few months or years of satisfactory growth and the break is smooth and clean.

Fig. 18. Development of an outgrowth at union, a symptom of graft incompatibility.

Types of graft incompatibility

Graft incompatibility is of three types 1) *Localized incompatibility* and 2). *Translocated incompatibility* and 3) *Virus induced incompatibility.*

1. Localized incompatibility: This type of incompatibility reactions apparently depend upon the actual contact between the stock and scion and localized in nature. This type of incompatibility can be overcome by insertion of mutually compatible interstock between the stock and scion, which separates the incompatible components of graftage. In this incompatible combination, the union structure is oftenly weak resulting in discontinuity in cambium and vascular tissues of stock and scion. It may result in poor translocation of the metabolites across the graft union. The external symptoms appear slowly depending upon the degree of anatomical differences at the graft union. Usually masses of undifferentiated parenchymatous tissues are commonly found at the graft union or even inclusion of bark tissue may develop, which may disrupt the normal vascular connection between the stock and scion. Localized incompatibility reactions are quite common in apple, when grafted on pear, and plum on cherry. An example of this category of incompatible reaction is that of Bartlett pear grafted on quince rootstock which is an incompatible combustion however with the insertion of compatible interstock, Old Home pear, the combination becomes compatible and satisfactory tree growth takes place.

2. **Translocated incompatibility:** Sometimes the incompatibility reaction is due to some labile influence, which can move across the graft union. Such type of incompatibility reaction can be corrected by insertion of a mutually compatible interstock. In this type, usually phloem degeneration takes place resulting in formation of brown line and necrotic area in the bark. It imposes restriction on the movement of carbohydrates through the graft union, with the accumulation above and reduction below. Most important example of this category of incompatibility is that of 'Hale's Early' peach grafted on 'Myrobalan B' plum rootstock. This forms a weak union in which distortion of tissues takes place with the accumulation of starch at the base of peach. If a mutually compatible interstock, 'Brompton' plum is inserted between these two, the incompatibility still exists, with the accumulation of starch at the base of 'Brompton' interstock. But if the same combinations are tried at cotyledonary stage, these are highly compatible indicating that some factors responsible for incompatibility reactions are not present at juvenile or nursery stage of the plants. Similarly, the combination of 'Non Pareil' almond on Mariana 2624 plum shows complete phloem breakdown, although the xylem tissues are normal. However, the 'Texas', another almond is highly compatible on 'Mariana 2624'. If a piece of 'Texas' almond is inserted as an interstock between 'Non Pareil' and 'Mariana 2624', bark disintegration occurs, resulting in incompatible graft union.

3. **Virus induced incompatibility**: Sometimes virus induced incompatibility cases are apparent and widespread. The failure of successful graft union can be due to pathogens. For example, sweet orange budded on sour orange in certain parts of the world showed incompatibility while in other regions, it was successful. After thorough studies, it was found that incompatibility was mainly due to viruses. The compatibility reactions change with the time also. For example, the pear cultivar 'Bristol Cross' when grafted on 'quince' in 1932 made good growth but required interstock 'Beurre Hardy' 30 years later to form an acceptable union. Such changes in plants may result from mutations or from latent viruses.

In certain cases delayed incompatibility symptoms were observed where in initial stage graft union was successful but later they showed incompatibility and usually clean breakage at union point was observed. For example, Black line in walnut.

Causes of graft incompatibility

Although, grafting is totally a physical phenomenon but compatibility and incompatibility reaction is the result of genetic differences between the stock and scion. However, the mechanism by which a particular case is expressed is not fully known. Many proposals or attempts have been made to explain the causes of incompatibility but the evidence supporting them are inadequate and conflicting. However, of the possible causes, structural, physiological, biochemical, incidence of diseases, insects or combination of all these factors may be responsible for incompatible reactions as explained briefly hereunder.

1. **Structural or anatomical reasons:** In some studies, it has been found that stock and scion may not differ structurally but abnormalities may be found due to the presence or development of parenchymatous cells at the graft union, preventing the formation of vascular continuity between the stock and scion. Similarly, from micro-spectrographic examination of the cell walls of incompatible graft union, it was found that the cell wall of the adjoining cells is lacking in lignin whereas in the compatible forms were lignified. Incompatibility reactions in plum, pear and peach are primarily due to structural abnormalities. Usually, development of undifferentiated tissues at the graft union causes mechanical weakness, resulting in phloem disintegration and uneven growth of graftage.

Cellular abnormalities in callus tissue seem also to be involved in the lack of cambial and vascular formation; thereby, avoiding the development of a strong union. In woody plants such as pear and quince grafts, an abnormal process of neocambium differentiation, which leads to cambium involution and lack of differentiation in to new vascular elements has been pointed out for incompatibility.

Plasmodesmeta are highly dynamic structures, which facilitate a unique pathway for symplastic communication between cells and constitute a potential pathway among cells in the graft bridge. In graft interfaces of incompatible heterografts, discontinuous-half plasmodesmeta have been observed in the graft unions between different types of cells. This indicates that plasmodesmata may contribute to graft failure due to misalignment of the graft partners, although they may not be unique to graft compatibility.

2. Physiological and biochemical reasons: Sometimes, there is insufficient supply of the necessary components either by stock or scion. Usually, in incompatible unions, the water supply is impeded. It may be due to failure of phloem or xylem continuity at the union resulting in root starvation and subsequent wilting and death of the scion. In compatible unions, translocation of sugars from scion to stock is normal. In many cases, there is accumulation of assimilates in the scions of incompatible combinations. For example, when certain pear cultivars are grafted on to

quince, a cyanogenic compound (prunasin) found in quince (not in pear) is translocated to pear phloem, which upon breakdown form hydrocyanic acid. This acid accumulates near the graft union resulting in tissue breakdown and lack of cambial activity and pronounced anatomical disturbances in the phloem and xylem of the union, hampering the movement of water and other materials. Synthe sis of flavanones can determine incompatibility in *Prunus,* such as prunasin and can be stimulated by hormones such as ABA and GA.

Different methods for an early detection of graft incompatibility have already been used, like isozyme analyses for peroxidases, presence of proteins in phloem, presence of acid phosphatase in *Prunus avium/ Prunus cerasus* grafts and phenol analyses. Graft compatibility occurs when the isozyme band pattern between scion and rootstock is similar. Moreover, the presence of isoperoxidases A and B in the graft union tissues may be used as an indicator to predict a compatible graft between pear and quince rootstock. Diverse phenolic compounds have been implicated in processes of division, development and differentiation into new tissues. In incompatible grafts, differentiation is not complete. The presence of phenols was generally associated with small cells in incompatible combinations, which did not form successful unions. Auxin (IAA) is involved in the development of compatible unions and induces the differentiation of vascular tissues. The activity of IAA oxidase and transport of IAA transport can be altered by naturally occuring substances such as phenols. Coumaric acid is an IAA antagonists or auxin–oxidase cofactors, but phloroglucinol is an IAA oxidase inhibitor. Starch accumulation above the scion or sites shortage on grafting interface might cause decay of phloem. Graft components with poor growth, not only leads to structural abnormalities in the site of scion, but also is correlated with irregularities in starch distribution.

3. **Nutritional deficiency:** Deficiency of certain nutrients may also result in incompatible unions. It has been found that when Jonathan apple is grafted on to EM-IX rootstock, the scion develops molybdenum deficiency. It may be due to the inability of EM-IX rootstock to absorb Mo in sufficient quantity to supply to the scion cultivar. However, Jonathan on other rootstocks does not show such deficiency symptoms. Similarly, deficiency of P, K and Mg in peach has been reported if grafted on to Myrobalan B plum rootstock, resulting in incompatibility symptoms.

4. **Presence of viruses:** Presence of latent viruses and mycoplasma like pathogens may result in failure of graft unions. Pear decline disease on Bartlett trees grafted on *Pyrus pyrifolia* rootstock is quite common due to the presence of viruses at the graft union. However, it never appears if *Pyrus communis* is used as rootstock. Further, the incompatibility reactions in citrus are primarily due to infection by viral diseases like tristeza, psorosis and xyloporosis etc. The failure of rough lemon rootstock for many sweet orange varieties in Punjab has been due to occurrence of

bud union crease a viral disease. Similarly, black line of walnuts, a delayed incompatibility reaction is believed to be due to a virus and not due to a rootstock failure.

Different Methods of grafting

Grafting may be accomplished by several methods, depending on the species, the age and size of the plant, the problem to be corrected and the purpose. Although, some methods are for general propagation purpose use, others are used to solve specific and specialized problems. Notwithstanding the methods, the principles are same for all types of grafting. There are two basic methods- one in which the scion is removed from its source and transferred onto another plant and a the second in which no detaching is done before the formation of graft union.

Attached methods of grafting or approach grafting

In approach or attached method of grafting, the scion is attached with the stock, when both the components are on their own root system. After the union takes place, the top of the stock above the graft union and the base of scion plant below the graft union are removed. The stock and scion are looped off gradually rather than giving a single cut at once. This method is often used by placing seedling plant growing in a pot or in a container under an established plant, which is to furnish the scion part of the new grafted plant. The various forms of approach grafting are described below:

i) **Inarching:** It is also known as simple approach grafting, simple inarching and embracing. In low rainfall areas, it should be done with the onset of rains while in regions of heavy rainfall, it should be done soon after the rainy season is over, provided temperature does not fall below 15^{0}C. Heavy rainfall, hot and dry period adversely affects its success. Selection of parent plant for taking the scion is an important factor for its success. For the actual inarching process, the stock is brought close to the scion. A slice of bark along with a thin piece and wood (about 4 to 6 cm in length and about 1-1.25 cm in thickness) at about 20cm above the ground level is removed from the stock with a sharp knife. A similar cut is made in the scion. Thus, the cambium layer of both stock and scion are exposed. These cuts are brought together and tied firmly with the help of a polythene strip (**Figure 19**). After the successful union, stock above and scion below the

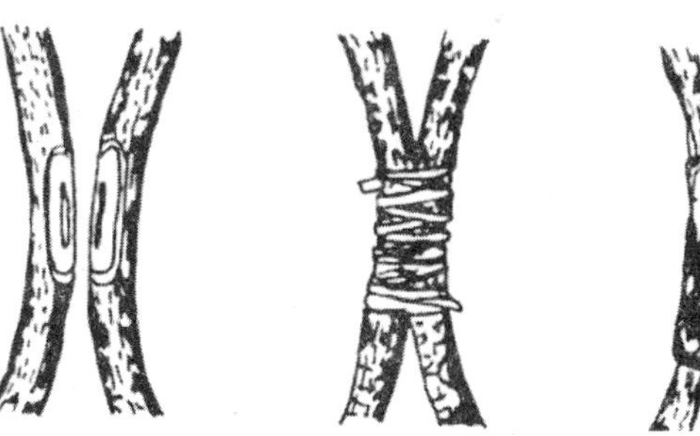

Fig. 19. Inarching approach grafting

graft union are looped off gradually. The scion plant should be healthy, vigorous and high yielding. Old plants selected as scion usually thrive for a shorter period. Inarching is also employed to rejuvenate the old, weakly growing, diseased or damaged tree to give a new lease of life, which is known, as *'adjuvant grafting'*. For this, a good number of seedlings are planted around the trunk of tree and when the seedlings get established, these are inarched to the old plant as discussed earlier. Increasing is commonly followed in mango, guava, loquat, sapota, aonla and tamarind. It can be performed during spring or rainy season or any other time of year when sap of both stock and scion flows freely.

ii) **Tongue approach grafting:** It is a modification of simple inarching and is generally practiced with thicker rootstock and scion. It is almost similar to splice approach grafting in which a second cut is made into the wood of stem about half way down to the first cut in a sloping direction pointing tongue upwards. A similar cut is made in the scion shoot. Both cuts are made in such a manner that one fits tightly into the other (**Figure 20**). Tongue grafting is little more complicated than inarching. However, tongue grafting provides better success because three surfaces of the cambium layers are involved in the union. Apple, pear and walnut are propagated by this method.

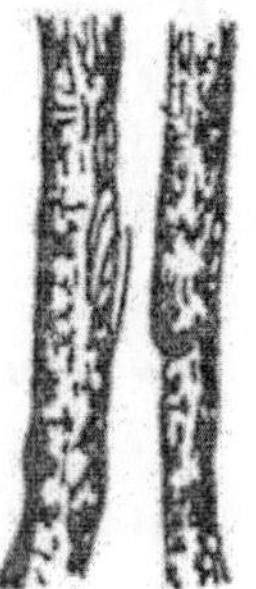

Fig. 20. Tongue approach grafting

iii) **Splice approach grafting:** In splice approach grafting, both stock and scion should be of same thickness. In this method, a splice of 4-5 cm bark and wood is removed both from stock and scion. The cut on both the partners should be of same size and should be smooth too. The cut surfaces of stock and scion are bound together with some suitable tying material and waxed properly. The stock above the union and scion below the union are cut after sometimes, when proper union has taken place.

iv) **Bridge grafting:** Bridge grafting is basically not a method of propagation but is a form of repair grafting in plants, which have been damaged either by frost, rodents or insects. It may, however, be kept in mind that bridge grafting is only helpful if the trunk is damaged but the root system of the plant is all right. In this method, first the damaged portion of the stock is cleaned and the irregular edges of the girdled area are cut evenly. The scion of desired variety is then inserted in such a way that it is attached at both upper and lower ends into the living bark. Similarly, it is important that scions be inserted right side up. After the scions are inserted, the cut surfaces must be thoroughly covered with grafting wax. The exposed injured wood must also be covered otherwise it may serve as an entry channel for decay organisms.

Detached methods of grafting

In this, the scion is completely severed from the parent plant during the process of union of stock and scion. The various methods of detached grafting are described briefly hereunder:

i) **Splice or whip grafting:** It is a very simple, and popular method of grafting. In this method, it is essential that both the stock and the scion should be of same diameter. About one-year-old rootstock is severed at a height of 23-26 cm from the ground level and a diagonal cut (3-4 cm long) is made at the distal end of the rootstock. A similar slanting cut is made on the proximal end of the scion and cut surfaces of both the rootstock and scion are bound together with polythene tape.

ii) **Tongue grafting:** This is a modified form of whip grafting. It differs from whip grafting in the sense that a tongue is added to the cut surface to provide better filling and rigidity. For this purpose, a shallow cut is given on both the stock and scion. When the tongue is prepared, the scion is slipped into the stock, so that they interlock easily. The graft is then tied and wrapped securely (Figure 21).

Fig. 21. Tongue grafting

iii) **Side grafting:** This method comprises the insertion of the scion base for about 2.5cm under the bark on side of the stock. In side grafting, the thickness of stock is more as compared to scion. It is one of the important method of propagation of mango (Plate 1). Two types of side grafting, that is stub grafting and side tongue grafting are commonly followed by the nurserymen. In this method, a slanting cut (2-4 cm long) is given half way deep on the rootstock, about 15 cm above the ground level. The scions are collected from current season's growth and two slanting cuts (one 2-4 cm long on one side and other 1 cm on the other side) are given for exposing cambium layer. The stock is made to bend away from its cut and the wedge shaped scion is inserted into the cut of stock and the shorter cut is covered by the bark of stock. The stock and scion are wrapped and tied with polyethythene strip tightly. It is commonly followed in mango. Side grafting differs with Veneer grafting in the manner that in case of veneer grafting only one side of the scion is sliced away in a sloping manner, whereas in side grafting, the scion is sliced on both sides of the lower portion in the form of a wedge.

iv) **Veneer grafting:** It resembles side grafting and is being increasingly used for propagation of many fruit plants. It is a simple method of propagation. It can be used in rootstocks of one year having a diameter of 1.0 to 1.5 cm. For veneer grafting 4-6 months old scion shoots with lush green leaves are selected usually, the terminal and next to terminal shoots are most ideal. The shoots are defoliated 7 to 10 days prior to grafting (depending upon the season) leaving the petiole attached. Defoliation of the shoots is not required under humid conditions. March to September (avoiding May to June) is ideal time for veneer grafting in northern India. The rootstock is prepared for veneer grafting by making a slanting cut (5 cm long), an oblique cut is then made at the base of the first cut so that a piece of wood along with bark is removed. The base of the scion wood is then fitted into the rootstock in such a manner that the cut surface including the cambium layers of scion and rootstock face each other. The rootstock and scion are tied together with polyethylene tape (**Figure 22**). When scion growth begins, the shoot of rootstock is removed above the graft union in one or two strokes. The technique is easy-to-operate, more economical and gives high degree of success. Besides, it is the most ideal method for establishing *in situ* orchards and top working of old unproductive orchards. Further, scion can be easily stored for 8-10 days at ambient temperature (25 to 35^{0}C) in moist sphagnum moss grass covered with polyethylene. The added advantage of this technique is that scion of desired variety can be procured from distant places. Mango is commercially propagated by veneer grafting in North India.

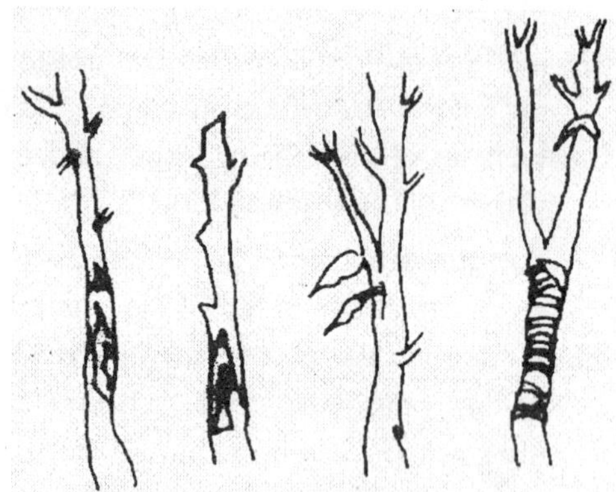

Fig. 22. Veneer grafting in mango.

v) **Cleft or wedge grafting:** It is one of the oldest methods of grafting and is employed in rootstocks having diameter greater than the scion. Rootstock with 5-7cm or more girth should be selected and cleft grafted after decapitating the stock 45cm above the ground level. The beheaded rootstock is split to about 5 cm deep through the center of stem with a knife and a mallet. When the knife is removed, a hard wooden wedge is inserted to keep it open for the subsequent insertion of the scion. Scion (15 to 20 cm) is used from a terminal shoot, which is more than three months old, and it is wedged securely (6-7 cm). The cleft of the scion is slipped into the spilt of the stock that has been forced open with a wooded wedge. It is usually done during the later part of the dormant season or just before the start of the active growth, as indicated by swelling of buds. Walnut, hazelnut, pecan nut and grape are propagated by this method. In addition to it, this method is commonly employed for rejuvenating the old orchards by top working.

vi) **Bark grafting**: The bark grafting is sometimes preferred to cleft grafting because it can be done without splitting the stub, thus preventing the entry of decay organism. It can only be done when the bark of the stock starts to slip in the spring. The scion used in bark grafting, should, however be dormant. In this method, the stock is first sawed off at a point, where bark is smooth. If the stock is thick, many scions can be inserted. For each scion, bark is split downward, about 5cm from the top end of stub. Scions of 10-12 cm long, containing 2-3 buds are collected from dormant wood and are prepared by giving a slanting cut (5 cm) downward along one side of the base. On second side, a small cut is given. The scion is, thus, inserted in the centre of split between the bark and wood of the stock. The longer cut of the scion is placed against the wood. After the scion has been set, all exposed portion of the stub and scion should be covered with a protective covering of melted wax (**Figure 23**).

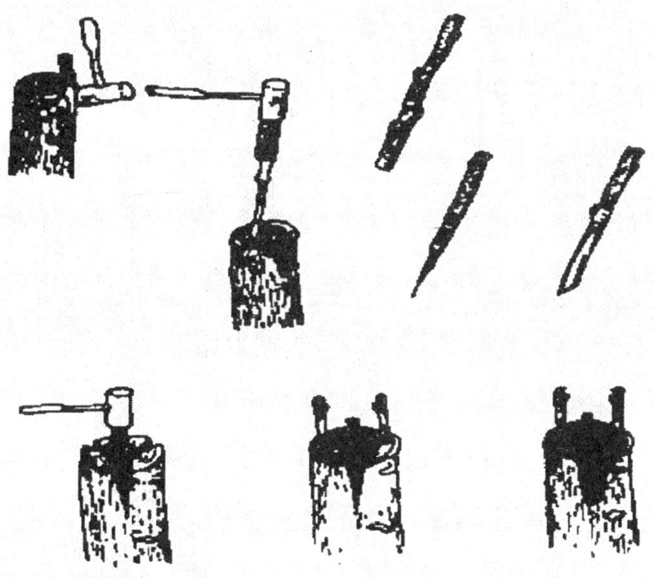

Fig. 23. Different steps involved in bark grafting

vii) Softwood grafting: In this technique, grafting is done with mature, pre-cured scion on the emerging soft coppery red shoot of rootstock. The leaves on the stocks must be retained while grafting, to get high success (**Figure 24**). Softwood grafting is successful in mango, avocado, sapota and tamarind. It is also used for *in situ* grafting, especially in the dry lands and areas experiencing hot weather with low precipitation where mortality of nursery raised grafts are very high.

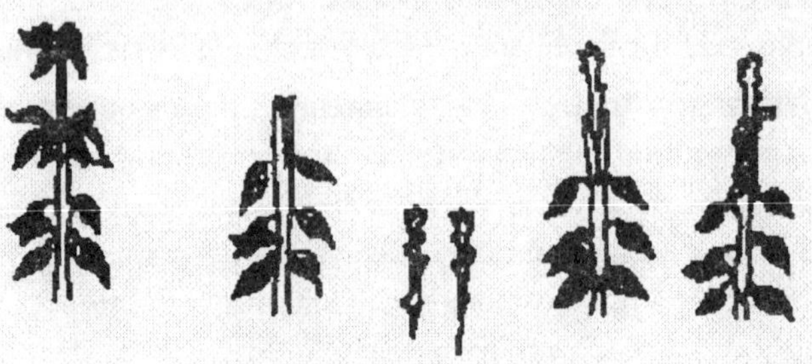

Fig. 24. Softwood grafting.

viii) Epicotyl grafting or stone grafting: In this method, germinating seeds of less than two weeks old are wedge or splice grafted with mature scion. Moderate temperature and high relative humidity are major factors related to success of epicotyl grafting. For splice grafting, germinating seeds less than 2 weeks old are used as rootstock. A 2-3 cm long slanting cut is made in the epicotyl with a matching cut on the proximal portion of the scion. In wedge grafting, the selected 2 weeks old seedlings are headed back by retaining 6-8 cm long stem. The longitudinal transverse cut running 4-6cm centrally down on the beheaded rootstock is made with the help of a sharp grafting knife. A wedge shaped cut, slanting from both sides, is made on the lower side of scion stick. The scion stick is then inserted into the saddle-like cut made on the rootstock and pressed properly so that cambium tissues of the rootstock and scion stick overlap each other. The union is then tied with the help of 150-gauge polyethylene strip (**Figure 25**). This technique has been commercialized for rapid multiplication of mango in Maharashtra, in general and Konkan region in particular.

Fig. 25. Epicotyl grafting.

ix) **Saddle grafting:** In saddle grafting, the rootstock is beheaded 25 cm above the ground level. Two upward sloping cuts are made on the side of the rootstock to form a 5-7 cm long wedge on its cut end. On the stem of the scion shoot, a tongue is made. The wedge of the rootstock is then fitted into the groove of the tongue and tied firmly.

x) **Notch grafting:** It is also called as inlaying. It is particularly useful in humid tropical areas. It is useful method when girth of the rootstock exceeds 8 cm. In this method, the rootstock is decapitated 45 cm above the ground level and V-shaped notches (four or five) are made at equal distance depending upon the thickness of rootstock, from the periphery of the cut surface. The proximal end of the scion is fitted into the notch of the rootstock and scions are sliced on to the notches made on them. Both stock and scion are tied firmly with polyethylene strip.

xi) **Double working:** More than two kinds of plants can be combined together in vertical arrangement. In addition to rootstock and scions, a third kind may be inserted in between called as interstock or intermediate stock. Thus, a double worked plant has two graft unions, one between rootstock and interstock and another between the interstock and scion (**Figure 26**). Double working is specially required for over coming graft incompatibility and for obtaining certain beneficial effects of interstock. The most exciting examples of double working is the propagation of Bartlett pears on Quince, a dwarfing rootstock by using a mutually compatible interstock such as 'Old Home' or 'Burre Hardy' Pear.

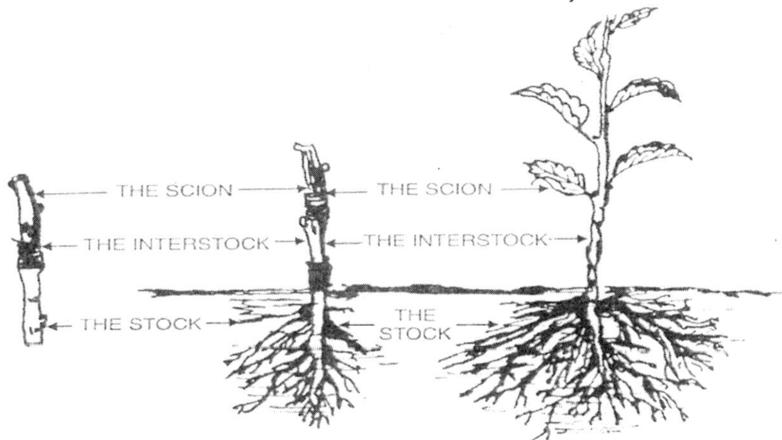

Fig. 26. Double working involving three different components.

xii) **Top working:** Top working/grafting is used primarily to convert an old or unproductive orchard of inferior variety with the desirable one. Top working should be done in spring; however, the exact period depends upon the fruit crop and method of grafting. The top working is an extreme type of severe pruning and one must decide in advance the number of scaffold branches to be retained, for maintaining a considerable balance between root and shoot system. The branches to be grafted should be well distributed all along the tree. Top working is most successful if grafting is done on scaffold branches not more than 10 cm in diameter. It is always better to retain some nurse branches while pruning, as these will be required for food preparation. Any of the grafting methods can be used for top working; however, cleft, side and veneer grafting are most commonly used. Mango and cashew nut are commercially top worked by side veneer grafting, while apple is top worked by cleft grafting (**Plate 2**).

xiii) Texas-method of Inlay grafting: Inlay grafting is one of the best and most popular systems of propagating pecans in U.S.A. It can be successfully used when other systems may not work because of heat, drought and wind. It has also been successfully used on walnuts, apples, pears, grapes, rabbit-eye blueberries and persimmons. Though this technique follows the basic fundamentals of the standard bark graft; however, it is different from bark grafting in the sense that it uses an inlay cut and employs an entirely new system of covering the graft and stock. The inlay occurs when two parallel cuts are made through the stock bark forming a scion inlay pattern on the stock. Aluminum foil is used as a stock cover, reflecting sunlight and reducing temperatures around the graft. The Foil is covered with polyethylene film to assure constant high relative humidity around the graft (**Figure 27**). This system not only results in a high percentage of growing grafts, but also easy-to-use.

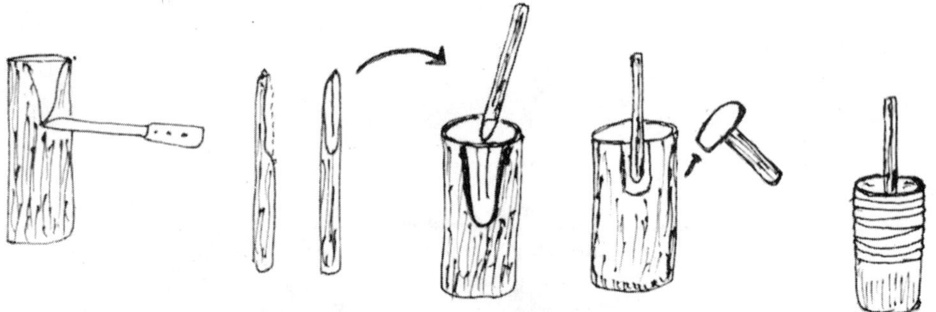

Fig. 27. Texas method inlay bark grafting.

xiv) Four-Flap Graft: The four-flap graft is ideal for small-caliper trees up to 1 inch in diameter before they are large enough for the Texas-method inlay bark graft. The four-flap grafts are most successful when the scion and rootstock pieces are near the same diameter. The best fit is obtained when the scion is slightly larger than the stock. Scion woods are collected from the desired tree when it is still dormant and stored until spring propagation time. Scion wood diameter should be 3/8 to 3/4 inch and in no case should exceed 1-inch diameter. The best time for grafting is when the rootstock is actively growing, the bark "slips" freely. Scion wood can be taken directly from cold storage and used immediately. Cut is made straight across the trunk or limb with sharp pruning shears at the point to graft. On the stock plant where the horizontal cut was made, four vertical are made, which are equally spaced and 1½ inches long. A special grafting tool 'Grantham' grafting tool is used to make the 4 vertical cuts. A smooth, straight piece of scion wood which is approximately of the same diameter as the stock and about 6 inches long is chosen. Cut is started about 1½ inches from the bottom end and knife is moved down. The four flaps of bark are moved down 1½ inches on the stock. This is why the graft is sometimes called the *banana graft*. The inner stock wood is removed and taking care that the 4 flaps on the stock are not damaged. Thereafter, insert the scion piece upright on the stock. Pull the four flaps on the stock up into place to cover the four cut surfaces on the scion. Place the bark strips of the scion in the cracks between the flaps of the stock. Wrap the cut flap areas with white budding tape. Coat the cut surface of the tip end of the scion with a protectant such as tree healing paint to prevent it from drying out. Loosely cover the taped area with a piece of household aluminum foil. This reflects sunlight and maintains a lower temperature at the graft union (**Figure 28**). Cut off the corner of a pint-size or quart-size polyethylene bag (freezer type) and carefully slip it down over the graft union with the scion protruding through a hole in the bag.

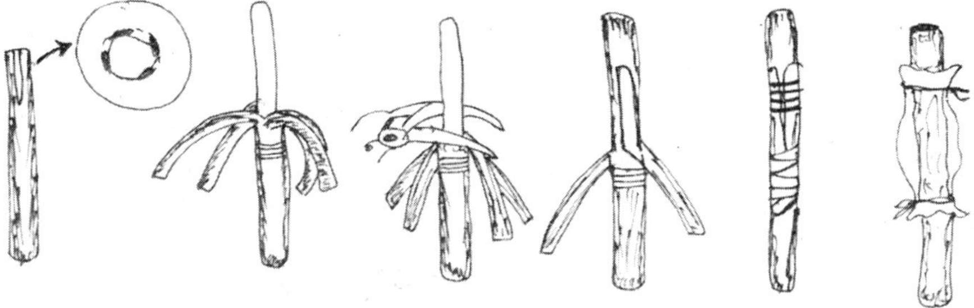

Fig. 28. Four flap or banana grafting.

xv) Green grafting: In some cases, such as grafting gooseberry, it is very difficult to graft wooded plant parts. Therefore, the two options to use are top grafting, when the stock was cut back or side grafting. Both cases use cleft or whip grafting methods. Top cleft grafting to produce a fruit tree requires a cut in the scion 20 cm long and also requires the removal of all leaves. Make two sloping cuts 2-5 cm long. Cut the top of the rootstock 20-40 cm above the surface. Make one straight cut across the center, the same length as the cut on the scion, firmly join the two parts and finally, tie and wax the graft. This technique is an effective and suitable tool for large-scale indexing of grapevine stock for viral contamination such as GFLV (Grape Fan Leaf Virus). Green-grafting has the advantage of being easy, having high success rates, and is also employed for overcoming the graft incompatibility sometimes experienced between distantly related *Vitis* species.

xvi) **Nurse root grafting:** Sometimes, it is desirable to have a stem cutting of a difficult-to-root species on its own roots. This can be done by root grafting, using the plant to be grown on its own roots as scion and a root of a compatible species as the rootstock. The procedure is the same as root grafting except that scion is a little longer than usual and the grafts are planted deep in the field with major portion of the scion below the soil. In it, scion rooting is promoted by using some auxins like IBA on the cuts made above the graft union before planting. After 2-3 seasons, the scion develops proper roots. The temporary nurse rootstock is then cut off and top is reduced in proportion to the root system. The rooted scion is replanted in the soil on its own roots. It is also a type of mound layering as the new scion shoot grows, soil is mounded up around to a height of 13-15 cm but terminal leaves are never covered. It is comparatively a cumbersome and laborious method of grafting because grafts are to be planted for 2-3 seasons for rooting and then these are to be dug up after the rootstock is cut off. Now several methods have been standardized to eliminate this cumbersome and laborious method. (a). By grafting rootstock on to scion in an inverted position. In this way, the union heals and the inverted stock sustains scion until roots are formed and stock will die as it won't receive food and water from the scion, thus leaving scion growing on its own roots. (b). Using incompatible rootstock, the scion will grow properly and the incompatible rootstock won't. (c). By wounding the scion above graft union by suitable wrapping material to girdle and cut off the rootstock.

xvii) **Nurse-seed grafting:** Nurse-seed grafting is used in woody species in which the seed germination is hypogeal i.e. the cotyledons remain below the ground in the seed coat and shoot tip appearing above the ground. In this method, the petioles of the cotyledons are cut off transversally just cut at the seed, leaving the cotyledons inside. A knifepoint is then inserted between the cut petioles for making an opening for inserting the scion. The scion is prepared from previous season's growth by making a wedge at the base. The scion is inserted between the cotyledons so that the exposed cuts of the scion and stock (seeds) make direct cambium contact. The grafts are wrapped carefully with suitable material like rubber strips or wax strings. These seed grafts are then placed in suitable rooting medium with graft union about 4 cm below the surface. The healing takes place within 2-3 weeks and roots arise from the petioles of cut cotyledons. Nurse seed grafting is done in early spring, which is successful in chestnut, avocado and camellia. A little modification of nurse-seed grafting has been very successful in propagating chestnuts, pecans and walnuts. In this method, the grafting is done on the seedling hypocotyl at the point of attachment of the cotyledons. The hypocotyl is cut longitudinally and split is made for insertion of the scion. The scion is prepared from the previous season's growth and is cut to a thin wedge. The scion is inserted into the split made in the hypocotyl, after which, the union is wrapped, firmly with a suitable wrapping material like rubber strips.

xviii) Stub Graft : In stub grafting (**Figure 29**), the main framework of the tree is retained. All suitably placed branches, 1 to 2 cm diameter at the base and with wide-angled crotches, are used for grafting. Other branches are removed completely before or during the grafting operations. The terminals of all branches are cut off at a point immediately above the uppermost stub graft. The greater the number of scions used, the less will be the interruption in fruiting. In each sector of the tree, scions should be placed and wounds covered, working from the top of the tree towards the base.

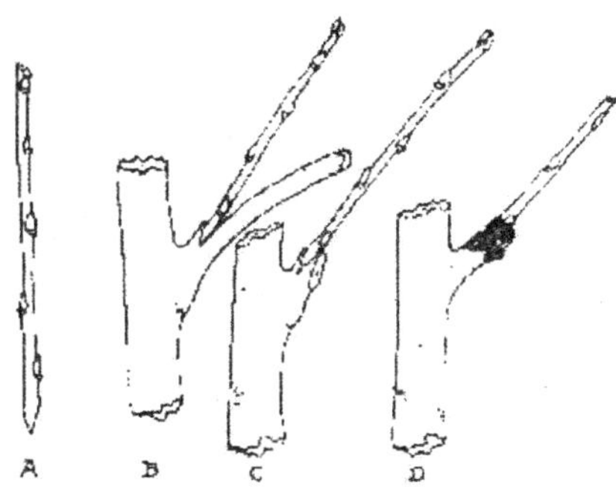

Fig. 29. The stub graft can be used effectively in frame-working. Here is the 6 to 8 bud scion (A) and the 3-step operation (B), (C) and (D).

xix) Inverted "L" : Like the side graft, this is a very useful method in frame working. Scions can be placed anywhere on the larger limbs. Make an inverted L-cut with an obtuse angle in the bark of the stock. Lift the bark from the wood and insert the bevelled end of a 6 or 8 bud scion beneath the bark (**Figure 30**). Using cigar box nails, or berry box nails tack the scion in place and cover all wounded tissue with grafting compound.

Fig. 30. The inverted "L" graft and the "awl" graft method. In both the cases, the bark of the stock is slit and the wedged scion slipped between the bark and the wood.

xx) Awl graft : This is more quickly and simply done (**Figure 30**) than the inverted "L" method. It is, in fact, probably the simplest of all grafting methods. Scions can be grafted anywhere on the larger limbs. A tool such as a bent screwdriver is used to make a slit in the bark. Insert the bevelled end of the scion into the slit and apply grafting compound. No taping or tacking is necessary. Be sure to insert the scions so as to avoid narrow-angled crotches.

xxi) Cutting-grafts: In this technique of grafting, a leafy scion is grafted on to a leafy, unrooted stem piece, which would be a future rootstock. After grafting, the components are placed in a rooting medium under intermittent mist for healing of graft and rooting of the stock. Splice grafting procedure is usually followed in cutting grafts. The slope of the cut is at 30° angle and the cut made is usually 1.5-2.0 cm long. The base of the stock is kept in rooting medium (IBA) for sometime and grafts are then placed in mist for healing process and rooting of the stock. After proper rooting and healing, the grafts are removed from the mist for hardening for 2-3 weeks. The grafts are usually ready for planting in pots or containers after two weeks hardening process. Initially, this technique was devised for studying the stock-scion physiology but now it is commercially used for propagation of citrus and clonal dwarfing rootstocks.

xxii) Root grafting: In this method, the roots are used as a stock for the grafting. The entire root system (whole root graft) or the pieces of the root (piece-root graft) may be used as a rootstock. In root grafting, whip method of grafting is used. It is usually performed indoors during late winter or early spring. The scion wood is collected well in advance like the rootstocks (roots), which are dug in late fall and kept under cool and moist conditions (35-40°F) until use. The term 'bench grafting' is also used for root grafting because it is practically done by propagators at benches. Apple, pear, plum, filbert, wisteria etc., are propagated by root grafting.In actual practice, the roots are cut into pieces of 7-15 cm. The scions of same length containing 2-4 lateral buds are used. The whip grafting is done in late winter or early spring and the grafts are bundled in groups of 50-100 and stored for callusing in damp sand or peat moss at a temperature of 7°C for about 2 months. The callus formation and healing process can be accelerated or retarded by temperature regulation. Ideal temperature for the healing process and callusing ranges between 7-21°C. Usually, the higher temperature accelerates and the lower temperature retards the healing process. Similarly, adequate aeration and moisture supply is equally important for healing process. After proper healing, the grafts are stored in a cool temperature (2-4°C) for breaking the rest period of the buds and to keep them dormant until planting. Afterwards, the grafts may be planted in the nursery during spring or early summer. These grafts should be planted in such a way that the grafts union remains well above the ground level. If the joint remains below the soil, it is likely that roots may develop in the scion, which is not desirable because one may not exploit the favourable influences (such as dwarfing or disease resistance etc.) of the rootstock on the scion cultivar.

xxiii) Crown grafting: Crown grafting is done at the stem-root transition region, which is called as the crown of the stock. It is specially performed in deciduous fruit plants in late winter to late summer. The scion of desired cultivar should be prepared from well-matured wood of previous season's growth. Different methods may be employed in crown grafting but whip, side or cleft grafting give good success, though it also depends upon the species and size of the rootstock. The grafting may be done just above, below or at the soil level. After grafting, the union must be wrapped tightly to hold both the components in position and avoid desiccation so that the healing of union takes place at an early date.

xxiv) Renewal grafting: It is often used on fruit trees when they have passed their normal productive life. New varieties are grafted on to the mature trees to produce a large fruiting tree in a very short time. It is almost similar to top working.

xxv) Micrografting: Grafting of tiny plant parts under aseptic and controlled environmental conditions is called as *micrografting*. In this method, small grafts are grown in closed containers until they grow large enough to be transferred to the open conditions. Micrografting has been used mostly in citrus, apple and plum. This technique was first developed by Spanish scientists, Navarro and his co-workers in late seventies for producing virus-tested plants of citrus. Micrografting is an effective technique for elimination of viruses, early diagnosis of grafting incompatibilities, rejuvenation of mature tissue, and bypassing the juvenile phase in fruit trees.

In shoot-tip grafting, embryos are excised from the rootstock seeds. These embryos are cultured in standard inorganic salt medium supplemented with one percent agar. After germination of embryo, the seedlings are removed and decapitated to 1 to 1.5 cm length. The cotyledons and the lateral buds are excised with a sharp blade. A 0.14 to 0.18 mm tip having three leaf primordia is used as a scion. This shoot tip eliminates the viruses. After this, an inverted "T" cut is given in the seedling rootstock by cutting 1 mm down the stem followed by a horizontal cut on the bottom. The excised shoot tip is then placed inside the flap, next to the cambium. Grafted plants are placed in a liquid medium. A filter paper bridge with a centre hole is made to support the stem. The cultures are kept in the light for 3-5 weeks for healing. When two expanded leaves appear on the scion, the grafted plant is pre-conditioned and hardened in the incubation chamber. Afterwards, it is transplanted in the field.The non-aseptic micrografting is also a promising procedure in some plant species for the production of virus free plants. In this method, propagation of small mini nursery plants is done by grafting a tiny seedling with scion of the same size. The minute grafts are then allowed to grow enough to have a viable plant. This technique is useful in tropical areas, where, nursery plants are sent to distant places either by ship or air because more number of plants can be accommodated in a smaller space.

* * *

7 Propagation by Budding: Principles and Practices

Budding is a form of grafting in which a single vegetative bud is taken of a plant (scion) and then inserted into the stem tissue of another plant (rootstock) so that the two unite together and eventually develop into a composite plant. Budding is also sometimes referred as bud grafting. It has many advantages over other methods of propagation;

- It is the best propagation method if propagating material is scarce.
- Budding is useful in plants, which release excessive wound gum (e.g. stone fruits) from injury carried to xylem or wood portion of stem at the time of grafting.
- Budding results into stronger union compared to grafting thus injury to the budded plants due to storms and strong winds is less.
- It is comparatively simple method of plant propagation than grafting and can be done by amateurs.
- It is a quick and efficient method of propagation as compared to grafting.

The success of budding depends upon bark's slipping ability. The period of year when plant is in its active phase of growth, cambium cells are actively dividing and the bark can be easily separated from the wood considered most suitable for budding. However, the factors like moisture stress, infection by insect-pests and diseases, defoliation or even low temperature may reduce the growth, resulting in tissue tightening and can interfere with the budding process. Among the different methods, chip budding can only be done when bark is not slipping.

Like grafting, rootstock has prominent place in budding. At the same time, selection of bud wood is equally important. Always use vegetative buds than flowering buds for budding. The vegetative buds are usually small and pointed while flower buds are large and plumpy.

Bud union

Like graft union, a series of changes take place in the formation of successful bud union also. In general, four stages *viz.*, pre-callus, callus, formation of cambial bridge and healing process etc. take place for the formation of successful bud union.

Pre-callus stage: This is called as the first stage of bud union formation. During this stage, there is no sign of callus formation and scion (bud) falls if the wrapping material is removed from the bud union. This stage lasts for 5 to 8 days after budding operation and is different for different species.

Callus formation: After 5th to 8th day of budding, the callus formation may take place in the bud union. The process of callus formation is fast in the firmly attached stock and scions and if the stock and scion are not in a close proximity, the callus formation and proliferation process is inhibited. The initiation of callus formation takes place either from stock, scion or both. Usually, living cells of recently formed xylem and phloem play active role in the callus formation but sometimes cells in the cortex or pith may also produce callus.

Formation of cambial bridge: The parenchymatous cells of the callus form a cambial bridge within 12-15 days after budding. Similarly, after the formation of a cambial bridge, de-differentiation of secondary xylem and phloem takes place after 36-48 days of budding. After that, there is partial movement of water and nutrients from soil to the stock and stock to the scion starts and establishes a better contact between the scion and stock. Thus, the bud union is complete within 6-7 weeks if conditions are favourable. However, if there is no better cambial contact with the stock and scion, the union may be weak and it may break after a few days of budding or broken by strong winds.

Healing of the bud union: After de-differentiation of secondary xylem and phloem of both stock and scion, there is formation of many new xylem and phloem after 6-8 months of budding. The callus unites completely with the scion pith and the union heals completely. Thus, like grafting, the success in the budding is also related directly with the proportion of live cells, environmental conditions, alignment of the cambial layer, uninjured cells near the proximity of both stock and scion. In general, better contact and alignment of the cambial cells of stock and scion helps in development and proliferation of callus. It prevents excessive accumulation of wound periderm and infiltration of resinous material, which hinders the healing process adversely.

Time of budding

The time of budding usually depends upon the availability of bud wood, active growth, cambium cells are actively dividing and the bark separates readily from wood. For most of the plants, this condition exists at three different times during the year as described below:

1. Spring budding

Budding is done in spring (March-April) when the active growth of the rootstock begins and bark separates easily from the wood. This period is very limited and budding should be completed soon. Bud wood is collected when it is still dormant and there is no evidence of bud swelling. Due to the combination of dormant bud and active growing rootstock, the bud sprouts and develops easily. Spring budding is commonly followed in citrus.

2. June budding

Budding is done in the early part of the growing season i.e., June and the inserted bud is forced into growth immediately during the same season. The bud wood used in June budding consists of current season's growth i.e., of new shoots, which have developed as growth started in the spring. By late May or June, these shoots usually have grown sufficiently to have a well-developed bud in the axil of each leaf. It is mostly done in stone fruits like peach, plum, apricot, cherry, nectarine, *ber* and *aonla* etc.

3. Fall budding

Actually, the budding is done in late summer rather than fall i.e., mid-June to mid-September. The bud sticks, consisting of current season's shoots are used for budding in fruit plants like *aonla*, mango, guava, *ber*, bael, jackfruit, stone fruits, walnut, chestnut and hazelnuts.

Preparation of bud stick

The leaves of the bud stick are trimmed, leaving only a short piece of the leaf stalk or petiole attached to the bud. It helps in handling. Middle and basal portion of buds of the shoots are considered best for budding. In certain species such as sweet cherry, buds on the lower portion of the shoot are flower buds, which of course, should be avoided. After few days of budding, if the leafstalk or petiole drops off cleanly, this is a good indication that bud is ready for sprouting.

Methods of budding

According to ease in performing the operation and percentage of success, the following methods of buddings are commercially advocated in propagating various horticultural crops:

i) **Shield or 'T' budding:** As the name indicates, 'shield' is the shape of the bud and 'T' is the shape of cut given on the rootstock for the operation. In this method, the bud is inserted in to the stock, 15-20 cm above the ground level and is tied with polythene strip (**Figure 31**). After the bud has sprouted, the stock is cut to about 10-15 cm above bud. When the bud has attained the length of 15-20 cm, the remaining portion of stock is cut right above the joint. Plants with thin bark, with

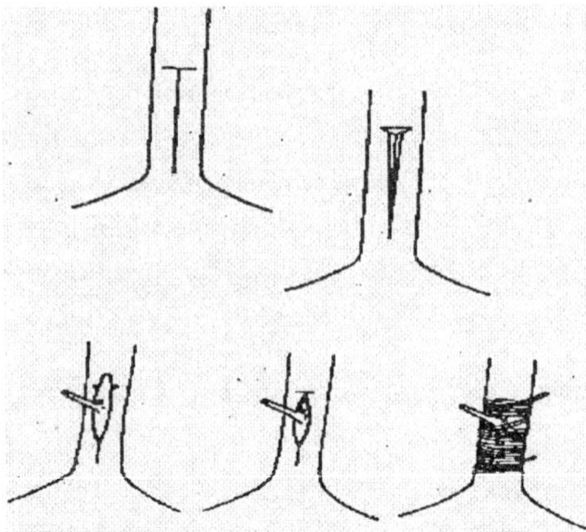

Fig. 31: T-budding

sufficient flow of the sap like apple, pear, peach, plum, and apricot, cherry, rose, ber and citrus are propagated by this method.

ii) **Inverted 'T' budding**: As the name indicates, the cut to be given on rootstock is reverse to that of 'T'. This kind of budding is widely used in areas of high rainfall or the species, which bleed badly during budding such as chestnuts. Inverted 'T' budding allows better drainage and better healing.

iii) **Patch budding:** In this method, a rectangular patch of a bark is removed completely from the stock and replaced with a patch of bark of the same size containing a bud of variety to be propagated. This kind of budding is slow and difficult to perform (**Figure 32**). However, it is successfully used in species having thick bark such as walnuts and pecans, where 'T' budding gives poor results. Mango, rubber plant, *aonla*, jackfruit and *jamun* are also propagated by this method.

iv) **I-budding:** The bud patch is cut like for patch budding i.e., in the form of square or rectangle. Then two transverse cuts are given on the smooth surface of the stock. These cuts are joined at their centres by a vertical cut to produce the 'I' figure. These two flaps of the bark are then lifted and the bud patch is placed beneath these flaps (**Figure 32**). The patch is tied as usual. I-budding is successful in plants species having thicker bark than that of bud stick.

v) **Ring or annular budding:** In this method, a complete ring of bark is removed from the stock and the bud stick of same size is inserted in it. New shoots arising from the heavily pruned plants of *ber*, peach and mulberry are capable of giving such buds for budding, which can be easily separated. In this method, the stock is completely girdled and if the bud fails to heal, the stock above the ring may eventually die (**Figure 32**).

vi) **Flute budding:** In flute budding, the bark patch is removed in such a way that it almost completely encircles the stock except with a narrow bark connection between the upper and lower cuts on the stock. A similar patch of bark is removed from bud stick containing a healthy bud and inserted in the stock. It should fit tightly (Figure 32).

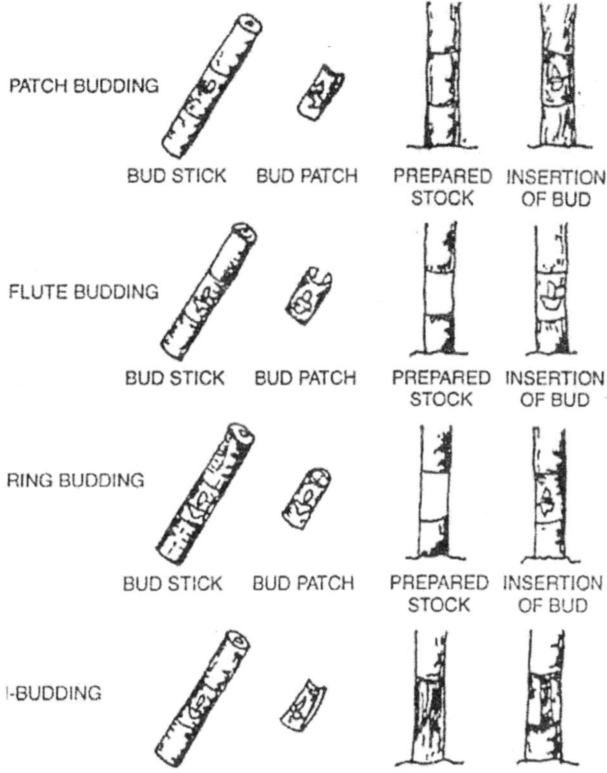

Fig. 32. Different budding methods.

vii) **Chip budding:** Chip budding can be used when the bark of the stock is not slipping. It may be due to poor sap flow or adverse growing conditions such as lack of water, defoliation or low temperature, which may lead to tightening of the bark and seriously, interfere with the budding operation. Under such circumstances, 'chip budding' is done in the beginning of February-March. Grape, apple and pear are propagated by this method. Besides, grape in dormant stage is also budded on nematode or phylloxera tolerant rootstock by this method.

viii) **Forkert budding:** In forkert budding, the stock is prepared by giving two vertical cuts and a transverse cut above the vertical cuts to join them. The bark is removed carefully along these cuts so that it hangs. The scion is prepared like patch budding and is fitted into the exposed portion of stock and the flap is drawn over the inserted bud patch. This is then tied with wrapping material. After successful growth of bud, the portion of stock above union is removed carefully.

ix) **Micro-budding:** This procedure is based on budding rootstocks that are only weeks-old, using small scion buds, and no taping to keep the bud in place. This technique enables the production of budded plants faster, at a smaller cost, and occupies only less space compared to the conventional system. The procedure works on several scion-rootstock combinations throughout the year, under greenhouse conditions. Micro-budded plants are used in biological indexing for tristeza, tatter leaf, and psorosis virus and exocortis viroid in citrus. It is similar to "T" budding except that the shield (bud piece) utilized in it is thin and tiny. For micro-budding operation, the petiole is cut off just above the bud and then bud is removed from the bud stick by a flap cut just underneath the bud. Thus, only the bud itself and a small piece of wood underneath the bud are utilized in micro-budding. In stock, an inverted "T" cut is made as is done in the inverted "T" budding and the tiny shield containing the bud is inserted in it and later tied with a thin plastic tape. The tape may be removed soon after (15-20 days) the healing has taken place.

x) **Skin budding:** Skin budding can be done in any season when plants are not in active growth and skin does not slip easily. It has been found to be very useful in citrus where it can be done at any time of the year. In skin budding, first of all, an oblong piece of 2.0 cm long bark is removed from the stock. The cut in the stock is given in such a way that it does not injure the wood. A bud of a desired cultivar is removed from a shoot of current season's growth. The bud is then placed on the stock containing a thin layer of bark along with it. It is then tied firmly with some wrapping material like polyethylene strip and sealed with melted wax. The stock is looped off above the bud union after the healing has taken place and bud has started taking active growth.

xi) **Top budding:** Like grafting, the old, declining and thrifty orchards of inferior cultivars can also be rejuvenated through top working, following one or the other method of budding. Thus, when budding is used to rejuvenate such orchards, the process is called as top budding. It can also be done on young vigorous plants having plenty of shoots. In older trees, the branches are headed back a year before the actual operation of budding to induce vigorous sprouts. Budding with any method can be done on the selected shoots with desirable cultivar. The number of buds to be utilized for top working depends upon the size of the tree but usually, 10-15 buds per tree are sufficient. "T" budding is usually practiced in thin barked species and patch budding in thick barked species. Further, it can be done in spring or rainy season, depending upon the climatic conditions of the locality and plant species to be rejuvenated. After budding, all other unbudded branches should be removed at the base. Similarly, the shoots arising in summer should also be removed regularly.

Bud wood selection and certification

It is now well-established fact that performance of an orchard lies on the selection and subsequent budding or grafting of a good bud-wood. It has been emphasized that the bud-sticks should always be collected from the plants of known history, productive, true-to-type plants, free from mutations and bud transmissible diseases. Bud-wood certification programmes need to be implemented to avoid the spread of diseases through bud wood. It is thought that the spread of citrus decline in India is mainly because of absence of any bud-wood certification programme, which has ultimately resulted in spread of viral diseases from one corner to the other corner of India. Thus, bud-wood certification will restrict the spread of diseases, thereby increasing production and productivity of the fruit crops. The bud-wood certification programmes, consist of preliminary selection, first flush inspection and registration of propagating material.

The preliminary selection of the mother plants is done on the basis of the trueness, vigour and productiveness of the plant. The plants must be older than 10 years. Such records are maintained by the orchardists. Preliminary inspection is done for quality of planting material particularly for visible disease symptoms and variation. Generally, plants of uniform age with uniform growth are preferred. In first flush inspection, parent tree is inspected for the presence of viruses. For this, field-indexing method is adopted to detect the presence of viruses. If the plants do not show any symptoms of viruses for 2-4 year, these are registered for further multiplication under supervision of the state agency.

* * *

8 Stionic Relationships

In the modern system of orcharding, mostly composite fruit plants are grown, comprising mainly of two parts i.e., the scion, which forms the top and the rootstock (stock), which produces the root system and in some cases a third part i.e., interstock, which is used for overcoming incompatibility between the scion and stock. All the components of a composite tree exert effects on each other; however, the effects of the stock on scion are more pronounced and have been harnessed widely in the modern fruit industry. Now there is hardly any fruit crop wherein no rootstock is used. The rootstock can't be changed during life-time of a plant so it makes it inevitable to pay more focused attention on it because its functions are long term and complex. Hence, before choosing a rootstock, it is essential to assess the proposed site in terms of soil and climatic factors because rootstock's performance is greatly affected by these factors. In this chapter, some key considerations related to rootstocks have been discussed.

Classification of rootstocks on basis of propagation

Rootstocks can be divided into two groups viz., seedling and clonal rootstocks on the basis of their method of propagation.

A. Seeding rootstocks

These rootstocks develop from the germinated seeds and have some advantages, like it is relatively simple and economical to raise them through seeds. Such rootstocks are usually well adapted to mass propagation. Moreover, in some cases, the root system tends to grow deeper and to be more firmly anchored than the vegetatively propagated rootstocks. Most of the seedling plants do not retain viruses dwelling in parent plants.

Seedling rootstocks have certain drawbacks, like genetic variation, which results in un-uniform growth and performance of scion cultivar (s) of the grafted plants. Seedling rootstocks are commonly used in mango, *ber*, pecan nut pistachio nut etc.

B. Clonal rootstocks

These rootstocks are those, which are multiplied through vegetative means either by stooling, layering, rooted cuttings or by aseptic tissue culture method. Such rootstocks have the same genetic make-up, which in turn result in uniform growth and performance of scion of grafted plants. In polyembryonic species like citrus, seedlings of nucellar origin are clonal and to great extent uniform in growth. The clonal rootstocks are more popular in fruit crops, like apple, pear, plum, cherry, grape and citrus.

Characteristics of an ideal rootstock

The characteristics of an ideal rootstock depend upon the soil, climate, specific needs of growers and market demand. However, the general, attributes by-and-large remain unchanged; therefore, an ideal rootstock should fulfill the following criteria Table 4.

Table 4. General attributes of the ideal rootstock.

Nursery requirements	Orchard requirements
• Ability to propagate reliably and cheaply. • Good graft compatibility. • Good nursery growth performance.	• Vigor control and uniformity of scion growth. • Induction of precocious, regular and abundant cropping in the scion. • Induction of good fruit size and quality. • Resistance/tolerance to pests/diseases. • Resistance/tolerance to drought or anaerobic soils. • Tolerance to soils with high pH. • Tolerance to severe winter cold. • Tolerance to viruses and phytoplasmas. • Freedom from abundant suckering or burrknots. • Low costs.

The performance of a graft or composite plant is the result of interaction of stock and scion. The relationship and interactions between two or more components of a graftage are quite variable and interesting. Some of these interactions have major horticultural value, whereas the others may not have any significance at all. These interactions are known to alter the size, growth, productivity, fruit quality and other attributes of different horticultural crops and thus are of commercial importance. The different interactions or the relationships occur among the different graft components are classified into three different categories:

i) Influence of rootstock on scion.

ii) Influence of scion cultivar on rootstock.

iii) Influence of interstock either on stock, scion or both.

1) Effects of rootstock on the scion

The effect of rootstock on the scion cultivar is one of the most significant influences. The rootstock influences the scion cultivar in many ways. The effects are variable, depending upon type of stock and scion. Some are:

a) **Anchorage :** The primary and most important function of rootstock is to provide anchorage to tree by growing deep into the soil. Many cultivars of different fruits have been selected for their desirable traits but somrtimes they do not have comparably suitable root system and hence require grafting on to rootstock to give rise to a satisfactory plant. Rootstocks with deeper root system avoid the need of staking and hence reduce the cost of cultivation. The dwarfing apple rootstocks like M_{27} and M_9, usually require staking particularly in early years due to brittle nature of their roots. Hence, these rootstocks could not prove to be beneficial to apple industry in Himachal Pradesh (India) because apple is grown in rainfed conditions whereas these rootstsocks perform well under irrigated conditions. Similarly, seedling plants owing to their deep root system are preferred as rootstocks in windy areas.

b) **Influence on size and growth habit:** In many fruit crops, rootstocks provide the principal method of controlling the excessive vigor of the scion cultivar. Control of tree size has become increasingly important in recent years as dwarf trees. Larger trees demand the use of ladders or expensive mechanical aids, and the cost per unit of quality fruit produced is higher than dwarf trees. The effect of rootstock to manage the plant vigour has been successfully utilized in several fruit species particularly temperate fruits such as apple, pear and stone fruts (Table 5-7). In India, Troyer Citrange rootstock has been recommended as a dwarfing, whereas Karna Khatta and Sohsarkar are semi-vigorous and vigorous rootstocks for Kinnow mandarin, respectively. Creeper and Vellaikolumban rootstocks have been reported to induce dwarfness in vigorous mango cultivars like Banganpalli or Alphonso.

c) **Influence on flowering and fruiting:** Rootstocks greatly influence the fruiting precocity, fruit bud formation, fruit set and yield of scion cultivars. Dwarfing rootstocks usually induce better yield precocity and abundance than more vigrous rootstocks. This is explained by the increased numbers and quality of floral buds produced on dwarfed trees. This, in turn, is likely to be partially attributable to the earlier cessation of shoot extension growth in the summer on trees on dwarfing rootstocks and the redirection of the trees' assimilates and nutrients toward the production of floral buds. Experiments at I.A.R.I, New Delhi have shown that the maturity of Kinnow mandarin can be staggered for 2½ to 3 months with the use of three different rootstocks viz. Troyer citrange, *Karna Khatta* and *Soh Sarkar*. The Kinnow fruits on Troyer citrange are ready to harvest by the end of November, which is considered to be very early, whereas, the maturity may be staggered up to mid-January by using other two rootstocks, viz. *Karna Khatta* and *Soh Sarkar*. Thus, the availability of Kinnow fruits can be extended invariably by the use of rootstocks, which is not only useful to the farmers but consumers as well.

Table 5. Rootstocks currently available or in advanced trials.

Sl. No.	Fruit Crop	Rootstock	Distinguisive influence on scion
1.	Mango	Totapuri Red Small	Dwarfing
		Vellai Kollumban	Dwarfing
		Rumani, Olour	Dwarfing
		Kurukkan	Salt tolerance
		Creeping	Dwarfing
2.	Grape	Dogridge	Resistance to *Phylloxera,* nematodes and salts.
		Salt creek	Resistance to salt and nematodes
		St George	Resistance to *Phylloxera* root louse
		Temple	Resistant against pierce's disease anthracnose and downy mildew
3.	Ber	*Zizyphus nummularia*	Dwarfing effect
4	Guava	Pusa Srijjan	Dwarfing effect on Allahabad Safeda scion cultivar
		Psidium pumilum	Used for the induction of dwarfing
5.	Custard apple	*Annona glabra*	Resistant to wet and damp soil conditions
6.	Fig	*Ficus glomerata*	Resistant to nematodes
7.	Avocado	Mexican race	Winter hardy
8.	Citrus	Flying Dragon	Most dwarfing, highly suitable for high density planting
		Trifoliate orange	Deciduous, cold hardy, dwarfing, resistant to nematodes, resistant to most viral diseases.
		Celopetra mandarin	Most salt tolerant citrus root stock
		Sweet orange	Resistant to tristeza, and exocortis
		Rangpur lime	Hardy rootstock, adaptable to various soil conditions and salt
		Severinia buxifolia	Salt resistant
		Rough lemon	Relatively tolerant to saline and calcereous soils
		Sour orange	Cold hardy, resistant to phytophthrora root rot but highly susceptible to triseza
		Citrus unshiu	Freeze tolerant

9.	Apple	M_9	Dwarfing effect and highly suitable for high density planting
		M_{27}	Ultra-dwarfing suitable for high density planting
		Malus sikkimensis	Induces precocity in bearing
		M_{26}	Better anchorage to scion cultivar
		M_7	Semi-dwarfing, stronger and deeper root system
		MM111 & MM106	Suitable for light sandy soils
		Northan spy	Wooly aphid resistant
		MM-series	Wooly aphid resistant
		EMLA series	Free from viruses
10.	Peach	Nemaguard	Resistant to nematode and crown gall
		GF-557	Nematode resistant
		GF-677	Drought tolerant, and high pH tolerant
		Prunus bessiyi	Dwarfing
		Montelor	Resistant to 'Fe' and 'Mg' deficiency
11.	Pear	Quince C	Semi vigorous
12.	Plum	Pixie and St. Julien	Dwarfing rootstocks
		Marian 2624	Resistant to nematodes, crown gall, cold hardy and tolerant to high soil moisture
		Myrobalan B	Resistant to bacterial gummosis
13.	Almond	GF-557 and GF-667	Tolerant to high soil pH
		Alnem 1	Resistant to rematode
14.	Apricot	*Prunus besseyi*	Imparts dwarfness to scion cultivar
		Haggith	Cold hardy
15.	Persimmon	*Diospyros virginiana*	Adapted to damp and heavy soils

Table 6. The current range of rootstocks for pear.

Quince rootstocks	*Pyrus* rootstocks
EMC	Pyrodwarf
Adams	BP 1
EM H	Fox 11 and 16
EM A	OH x F series
Sydo	
BA 29	

Table 7. The current range of rootstocks for cherries.

Species	Nature/ origin
Prunus avium (Mazzard)	Seedling or clonal (e.g. F 12/1)
Prunus mahaleb (St. Lucies)	Seedling or clonal (e.g. S.L.64)
Prunus ceresus (Sor cherries)	All clonal Table (Edabriz) Weiroot series (e.g. 53, 72, 158)
Prunus hybrids	All clonal Colt Max x Ma 14 Giesla series e.g. 5, 6 PHL series Inmil Damil

d) **Influence on fruit size and quality:** Rootstocks also influence the fruit characteristics of the scion cultivar invariably. Quince rootstocks often produce better fruit size and quality of pears than when seedling or clonal *Pyrus communis* rootstocks are used. The dwarfing apple rootstocks though do not influence fruit size but colour development in the scion cultivar is much better as compared to standard or seedling rootstocks. In citrus, the striking effects of the rootstock appear in fruit characteristics of the scion variety. Sour orange rootstock produces thin skinned, juicy and excellent quality fruits of sweet orange, tangerine and grapefruit. Rough lemon usually produces thick, coarse skinned fruits of inferior quality. Troyer citrange rootstock usually reduces fruit size in Kinnow but Karna Khatta and Soh Sarkar improves it. Rootstocks also affect the storage quality of the fruits. For example, the fruits produced on the dwarfing apple rootstocks have poor storage life as compared to those produced on standard rootstocks.

e) **Influence on insect-pests and disease resistance:** Dwarfing apple rootstocks are susceptible to wooly apple aphid and crown gall but Malling-Merton (MM) series of rootstocks (MM_{104}, MM_{106}, MM_{109}, MM_{111} etc) impart resistance against woolly aphids. French pear rootstock imparts resistance against fire blight disease to the scion cultivar. For example, Old Home pear is resistant to pear decline. Further, Mahaleb rootstock of cherry imparts resistance against buckskin virus. Similarly, a considerable variability exists in citrus rootstocks in their capacity to induce resistance against many diseases. For example, sour orange is highly susceptible to tristeza virus, whereas rough lemon and Cleopatra imparts resistance against it. *Citrus macrophylla* is a very good rootstock for lemons because it is known to impart resistance against gummosis, the most serious disease of lemons. *Poncirus trifoliata* rootstock imparts resistance against trisreza virus and root rot disease.

f) **Winter hardiness:** Some rootstocks impart resistance to scion cultivar for cold tolerance. Apple rootstocks, like EM-IX, EM XVII and Alnarp 2 have been found to be resistant to winter injury. EM-IX is resistant to early winter frost, though susceptible to late winter frost. Among citrus rootstocks, trifoliate orange is considered as the hardiest rootstock to low temperature followed by sour orange and sweet orange.

g) **Other influences:** Different rootstocks have been recommended for different fruit crops and even for different cultivars of the same species. Similarly, one rootstock recommended for one cultivar in a particular locality may not be suitable for same cultivar in the different locality. Thus, selection of a proper rootstock as per scientific recommendation for different regions having different soil and climatic condition is very important, because the success or failure of a particular fruit crop in that particular locality would be decided by it. For example, Jatti Khatti citrus rootstock performs well in semi-arid zones of Punjab and Rajasthan, whereas Karna Khatta is suitable in Indo-Gangetic plains of Uttar Pradesh but not the others. Myrobalam plum is highly suitable rootstock in regions of excessive soil moisture, whereas almond is most susceptible to wet conditions. Sour orange, Rangpur lime and Cleopatra mandarins are found to be resistant rootstocks for higher salt concentrations. In general, Rangpur lime has shown wider soil adaptability as compared to other citrus rootstocks. In mango, Kurukkan performs well in soils having higher salt concentrations.

2) Influence of scion cultivar on rootstock

It is not only the stock, which influences the scion, but the top (scion) can also influence the stock in many ways, though examples are very less but not the least. Influence on the vigour is the major effect of the scion on the stock. If a vigorous cultivar is grafted on a weak rootstock, the growth of the stock is also stimulated. Further, if a dwarf or a weak scion is grafted on the vigorous rootstock, the growth of the stock is also reduced. Similarly, the size, nature and form of the root system of the stock are also influenced by the scion cultivar. If the rough lemon seedlings are grafted or budded with Malta and Santra scions, a more fibrous root system is developed as compared to the non-budded plants. Similarly, Red Astrachan apple scion produces a fibrous root system with a few tap roots, while Olden Burg scion produces two or three prolonged deep tap roots but not fibrous root system. A strong scion seems also to cause the roots to be more resistant to wet, poorly aerated soil than the same roots would be if the scion were of a weaker growing variety.

3) Influence of interstock

The ability of certain dwarfing rootstocks, used as interstock between vigorous top for producing a dwarfing and early bearing trees is in practice. Paradise apple is widely used as interstock between scion cultivar and Crab apple rootstock, for inducing precocity in bearing. Similarly, use of Malling IX dwarfing rootstock as interstock is a commercially viable technology. In mango, Anupam has been recommended as interstock for Amrapali, grafted on Mallika rootstock, to induce dwarfness, precocity in bearing and improvement in fruit quality. However, sometimes interstock may not result in favourable or desirable influences.

Mechanism of dwarfing

Several rootstocks control vigour of scion cultivar. However, it is still a matter of great confusion as to how dwarfism is created by rootstock. It is still not completely understood in the light of available evidences, despite the fact that such dwarfing rootstocks are in vogue in apples and pears for centuries. The physiological mechanisms of vigour reduction by rootstocks or inter-stems are also not clearly comprehended. It has been proposed that dwarfing might be caused by water supply restrictions to the scion induced by anatomical characteristics of the rootstock. On the other hand, others suggested that dwarfing was caused by a reduction of solutes transported to the scion through the rootstocks. A long series of studies reported possible mechanisms related to the production and translocation of hormones in the plant. However, exactly how hormones act in the dwarfing process is not well understood. It has also been reported that in some cases dwarfing might be caused by partial incompatibility between the scion and the rootstock, which may alter the transport of minerals and hormones. It has also been suggested that the graft union may play a central role in limiting water transport and thus induce scion dwarfing.

***

9 Propagation by Layering: Principles and Practices

Layering is one of the oldest techniques to propagate many fruit plants. It was used extensively in European nurseries from the 18th to the early 20th century for propagating woody shrubs and tree species. Layering is a reliable means of plant propagation and is adopted for species, which are difficult or impossible to root from the cuttings or other vegetative means. Layering is a form of vegetative propagation where shoots are forced to form adventitious roots while still attached to the mother plant. Stems of the mother plant are covered in a growing medium in various ways to exclude light, increase the moisture level, and stimulate root growth. Once roots are formed, the new plant (layer) can be dug or cut from the mother plant and transplanted to the desired location. Layering does not lend itself to the production of large number of plants and therefore nurserymen use it only for difficult-to-root plant species. It is a natural means of propagation in black raspberries and trailing blackberries, or it may be induced artificially in many plant species. In general, better rooting in the layers can be obtained by ringing or wounding, etiolation (absence of light) and/or use of rooting hormones (IBA, NAA) and by providing favourable environmental conditions (temperature and humidity).

Advantages

Layering has certain advantages over other methods of propagation as under;

- It is an effective means of propagating species that usually do not root easily by cuttings as in mango, kumquat, filberts and litchi etc.
- It is the best method of propagation of plants, which reproduce naturally by layering e.g. blackberries and black raspberries.
- It does not require precise control on water, relative humidity or temperature as is required for other methods of propagation.
- It is easy to perform and does not require much facility.
- It is possible to produce large sized plant with layering within a short time.
- Layering is useful for producing relatively a smaller number of plants of good size with minimum propagation facilities.

- Allows propagation to be performed on-site.
- Natural accumulation of photosynthates and hormones due to girdling, incision, or bending.

Disadvantages

Layering has certain disadvantages, as under;

- It is a costlier technique in areas, where labour availability is a problem.
- It is not possible to produce large number of plants within a short time.
- The plants produced through layering have usually small and brittle roots.
- In layering, the beneficial effects of rootstocks on the scion cultivar can't be exploited.
- The mortality rate in layers (particularly air layers) is usually high.

Factors affecting the success by layering

Many factors affect the root initiation and their development in the layers. Some of the important factors are discussed below:

1. **Nutrition:** The stem remains attached to the plant during the process of rooting thus the nutrients and water are continuously supplied through the intact xylem, which helps in better rooting. Thus, nutritive condition of mother plant plays pivotal role in success by layering.
2. **Stem treatments:** Various manipulations like bending of the shoot, notching, girdling, wound etc., are made in the stem, which cause interruption in the downward translocation of organic materials like carbohydrates, auxin, etc. from the leaves and the growing shoot tips. This material is accumulated near the point of treatment, which helps in better rooting of the layers.
3. **Etiolation or exclusion of light:** Exclusion of light (etiolation) has strong influence on rooting of layers. The greatest stimulus to root induction results when the initially developing shoots are continuously covered with rooting medium as in case of mound/stooling and trench layering. In etiolation treatment, the base of the layered shoot (approximately 2.5 cm) is never exposed to light.
4. **Physiological conditioning:** Timing of layering is associated with the movement of carbohydrates and other substances. The rooting is better in layers if done in the end of the seasonal cycle of the growth.
5. **Rejuvenation:** Heading back/cutting back of the old plants encourages new shoot growth as in case of mound/ stool or trench layering. These shoots usually root better as compared to older shoots.
6. **Treatment with growth regulators:** Treatment of the wounded part of layers with IBA or NAA helps in better rooting. The concentration of these hormones

usually depends on the plant species, type of layering, age of the shoot and environmental conditions.

7. **Environmental conditions:** The environmental conditions, specially temperature and relative humidity exert greatest influence in rooting of layers. Profuse rooting takes place in air layers in high humidity. Similarly, rooting is better in stooling if mounds are kept wet throughout.

Layering types

Different types of layering techniques are used for propagating different horticultural and forestry plants depending upon the plant species, facilities and the environmental conditions. A brief description of these methods is given hereunder:

i) **Tip layering:** It is commonly followed in blackberries, raspberries and dewberries in which rooting takes place near the tip of current season's shoot. The stem of these plants completes its life in two years. During first year, vegetative growth takes place, while in the 2nd year fruiting takes place. After harvesting the fruits, the plants are heavily pruned, which gives rise to a number of lateral shoots. The tips of these shoots are buried 5 to 10 cm deep in the soil. Rooting in the buried shoots takes place within a month or so. Rooted layers are detached and planted in soil during spring.

ii) **Simple layering:** Flexible branches of plant are bent over to ground in early spring. A second bend is made in branch at a short distance from the tip, which is covered with soil and held in place with wire or wood stakes. This portion is sometimes injured or wounded either by notching, wiring or girdling to stimulate rooting. Filberts, guava, litchi, lime, lemon, etc., are propagated by simple layering (Figure 33).

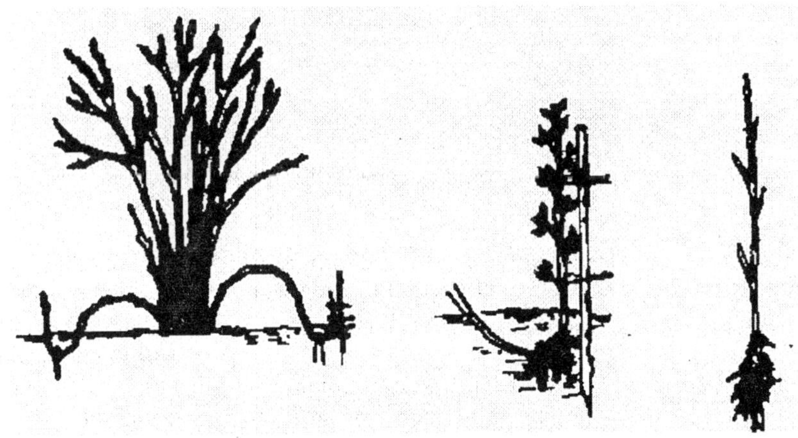

Fig. 33. Simple layering

iii) **Serpentine or compound layering:** It is modification of simple layering in which one-year-old branch is alternatively covered and exposed. The stem is girdled at different points in the underground part. The exposed part of stem should have at least one bud to develop a new shoot. After rooting, the sections are cut and planted in the field. In this way, many new plants can be made from one branch. Muscadine grape is commercially propagated by this method. Ornamental vines like *Wisteria* and *Clematis* are also propagated by serpentine layering.

iv) **Air layering (Goottee/Marcottage):** In this method, roots are formed in the aerial part of the plant. A branch/shoot 30-45 cm long and 1-1.25 cm thick, is selected and 2 -3 cm wide ring of bark is removed with a sharp knife. The exposed cambium is rubbed off with back of knife blade. The rooting hormone (IBA) is applied to upper part of cut. The moist rooting medium (moss grass) is wrapped with the help of a small transparent polyethylene strip (200-300 gauge) (**Figure 34**). Rooting in air layers generally commences within 25-30 days and layers are ready for transplanting within 3 months. This method is commonly known as gootee or marcottage or Chinese layering. Many fruit plants like litchi, Kagzi-lime, jackfruit, guava, fig, croton and cashewnut are propagated through air layering. February-March and June-July are considered as the most ideal months for air layering.

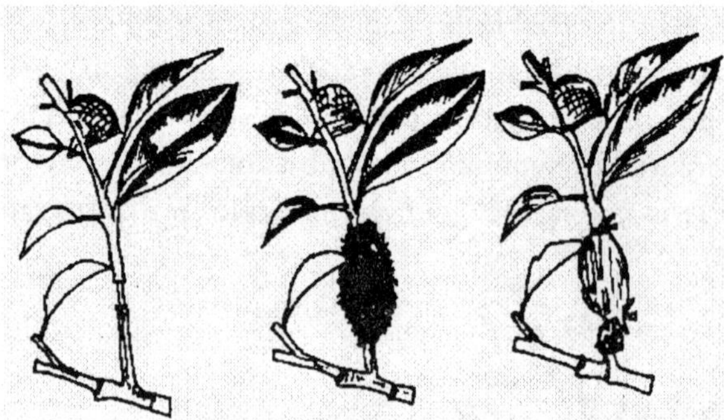

Fig. 34. Air layering.

v) **Mound layering or stooling:** In stooling, the plant is headed back to 15cm above ground level during dormant season. The new sprouts will arise within 2 months. These sprouts are then girdled near base and rooting hormone (IBA), prepared in lanolin paste, is applied to the upper portion of cut. The concentration of rooting hormone varies from plant-to-plant but in general, 3,000-5,000 ppm is most commonly used. These shoots are left as such for two days for proper absorption of rooting hormone (IBA) before they are covered with moist soil (**Figure 35**). Care must be taken to keep the soil heaps moist all the times as it facilitates rooting. The rooting of shoots is observed with 20-30 days. After 2 months, the rooted stools are separated from mother plants and planted in nursery. Apple (**Plate 3**), guava, quince, currants and gooseberries are commercially propagated by this method. However, this method is also advocated in other fruits like plum, cherry, hazelnut, pecan nut, mango, jackfruit, litchi, etc., and for production of uniform rootstocks.

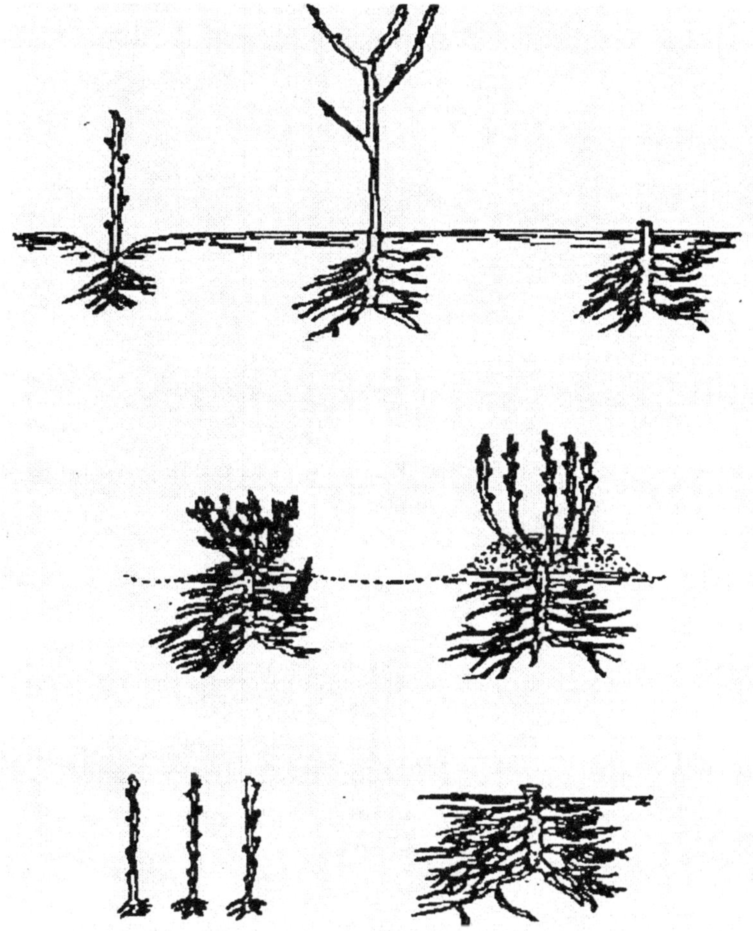

Fig. 35. Mound layering.

vi) Trench layering: In trench layering, a plant or branch of a plant is grown in a horizontal position in the base of trench and the trench is then filled with soil around the new shoots. Roots are developed at the base of new shoots. It is commonly used in woody fruit plants, which are difficult-to-propagate by mound layering. Vigorous rootstocks of apple like M_{16} and M_{25} and walnut are usually propagated by trench layering.

vii) Herbaceous layering: In herbaceous layering, very young shoots of plants (especially carnation) are used for producing new plants. The layering of border carnations is usually regarded as a skilled operation but it is not really difficult. In this layering, first of all, sandy compost is placed around the plant to be layered. Young shoots are selected and their lower leaves are removed. Now, a slit near the base, about halfway through the stem and in an upward direction is made. The stem is then bent so as to open the slit and at this point is pegged firmly into the soil and covered with an inch or so with compost. The layers will root by autumn and then be severed and potted.

viii) Pot layering: In south India, mango and sapota is propagated by pot layering.

* * *

10 Propagation by Specialized Vegetative Structures

Some plants develop certain natural specialized vegetative structures or organs, like bulbs, corms, tubers, tuberous roots, rhizomes and pseudobulbs etc. These vegetative structures are primarily function as the storage house of food, water use, which are utilized by the plants under adverse conditions of the environment. Plant belonging to this category are usually herbaceous perennials, whose shoot die at the end of growing season but the fleshy vegetative structure usually don't die and remain in the soil, which putforth new vegetative growth in the next season. These vegetative structures can also be used for raising new plants having characteristics of the mother plant. Propagators consider that propagation through specialized vegetative structures is the simplest and easiest method of propagation. Moreover, it is one of the most successful methods of propagation for a beginner. A brief description on propagation techniques used for species with specialized organs is below:

Tubers

It is a modified under ground stem, developed as a result of accumulation of reserve food material. The tuber has all characteristics of a stem having nodes, internodes, lateral buds and terminal buds. The eyes of a tuber are spirally arranged, consisting of one or more small buds, subtended by a leaf scar. Usually, one bud in the eye sprouts and suppresses the growth of the others; however, removal of the first sprout permits other buds to sprout. Potato, Jerusalem artichoke, dahlia (**Figure 36**) and caladium are some examples of tuber. Propagation by tubers is usually done either by planting whole or cutting the tuber into pieces with at least one bud or eye.

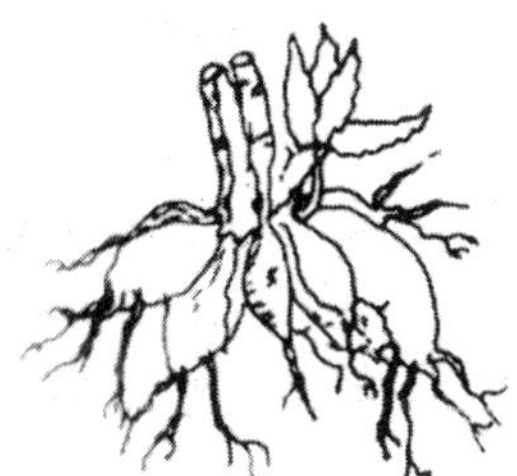

Fig. 36. Tubers of Dahlia.

Bulb

A bulb may be defined as a specialized organ having a short, fleshy vertical stem axis with growing points or flower primordium at its apex, enclosed by thick fleshy scales. Bulbs are of two types, i.e., **tunicate** and **non-tunicate**. Tunicate bulbs have dry and membranous outer bulb scales (e.g. onion, tuberose, daffodil and tulip). These membranous coverings provide protection from drying and mechanical injury. In non-tunicate bulbs, the dry coverings are loosely arranged like petals of a flower overlapping each other. e.g. garlic and lilies (**Figure 37**).

Fig. 37. Bulbs of lily.

Pseudobulbs

A pseudobulb (false bulb) is an enlarged fleshy stem with several nodes. It is produced by some horticultural plants (mostly orchids), mainly for the storage of food. The structure of pseudobulb varies with the species. The plants producing pseudobulbs are usually propagated through offshoots, division or from black and green bulbs produced by them. In few orchids, like *Dendrobium*, the offshoots are developed at the nodes of long slender pseudobulbs, from which the roots develop. The rooted shoots are cut from the mother bulb and planted in the nursery. Some orchids, like *Cattleya, Laelia, Miltonia* etc., are multiplied commercially through the division of pseudobulbs.

Bulbils

In some species, bulblets may form in the leaf axil either on the underground portion or on the aerial portion of bulb (stem). These aerial bulblets are called as bulbils. Garlic and some species of lilium like *Lilium bulbiferum* and *L. sulphureun*, produce bulbils and are propagated through it.

Corms

It is a short, solid, much swollen underground stem, with one or several buds near the top, a tuft of leaves at the upper side and a ring of thick fibrous roots around the base. Corm has distinct nodes and internodes and is covered by a thin membranous scale leaves. The typical examples of corm are gladiolus, crocus and water chestnut. Cormous plants are usually propagated through new corms or comlets, which are produced every year such as gladiolus (**Figure 38**).

Fig. 38. Corms of Gladioli.

Rhizome

It is a modified stem structure in which the main axis of the plant grows horizontally at or just below the surface of the ground. Banana is a typical example of rhizome. For propagation, each rhizome is cut into pieces in such a way that each piece contains atleast one lateral bud or eye.

Tuberous roots and stems

In some herbaceous plants, there is massive enlargement of the secondary roots. No nodes and internodes are present and buds are produced only on the crown of the stem as in dahlia and sweet potato. Tuberous stems are produced by excessive enlargement of the hypocotyl of seedling plant as in cyclamen.

Runners

It is a specialized long slender stem, which is produced from the leaf axil at the crown of plant and prostrate horizontally. The roots appear at one of the nodes having contact with soil. After root formation in the new plant, the contact with the mother plant is automatically detached and the new plant can be separated and planted. Strawberry is a typical example, which is commercially propagated through runners (**Figure 39**). Doob grass (*Cynodon dactylon)* and *Oxalis corniculata* are two other plants propagated through runners.

Fig. 39. Runner production in strawberry.

Suckers

A sucker is a shoot, which arises on a plant below ground. However, in practice, shoots arising from the vicinity of the crown of the plant, are also referred to as suckers. Suckers are separated from mother plant and are either planted in the nursery for rooting or directly in the field. Pineapple is usually propagated through suckers. In banana, two types of suckers are produced – water suckers and sword sucker. Water suckers are broad leaved, whereas sword suckers are pointed and have sword shape. For propagation purpose, sword suckers are preferred over the water suckers.

Offsets (Offshoots)

An offset is a young plant produced laterally by the parent plant or branch that develops from the base of main stem in some plants, which can easily be detached from it. The term offset is generally applied to a shortened, thickened stem of rosette like appearance. It is also applied to lateral shoots arising on the stems of monocotyledonous fruit plants, like date palm, pineapple or banana. Bromeliads produce offsets around the old flowering shoot. Similarly, some bulbs also produce offsets freely. Many plants are propagated by means of offsets. For example, fruit plants like date palm, pineapple, banana, and some succulents and cacti, and bulbs like tulip, daffodil, *Iris,* lily, hyacinth, amaryllis etc., are chiefly propagated through offsets.

Crown

It designates the above ground part of a plant from which new shoots are produced. In perennial plants and shrubs, the transition zones between the stem and the root, whereas in herbaceous perennials, the part from which new shoots arise is basically the crown. Some plants produce extensive crowns and others only one. Because of simplicity and reliability, crown division is an important method of propagation of many herbaceous perennials. In this method, the crown is divided in spring with the help of a sharp knife. The cut pieces are planted in pots or nursery bed immediately after preparation. Sansevieria, day lily and Shasta daisy are propagated through crown division.

Slips

Slips are the shoots just arising below the crown but above the ground. Pineapple is commercially propagated through slips.

Stolons

Stolon is a modified stem, which grows horizontally above the ground and produce a sprawling mass of stems along the ground. Stolon formation is common in Bermuda grass (*Cynodon* dactylis), mint (*Mentha*), *Ajuga,* dogwood (*Cornus stolonifera*). These plant species are sometimes involved in tuberization also. The stolon is a naturally occurring layering in which individual layer can be cut from the parent plant and planted in the nursery.

Tuberacles

Some plants like *Begonia* and *Dioscoria batatas* produce small aerial tubers in the axil of leaves called tuberacles. These tubercles are removed in fall and stored for 3-4 months in winter and planted in spring. Shoot and root emergence takes place within short time but tuber formation is induced under short day conditions.

* * *

11 *In Vitro* Plant Propagation: Principles and Practices

Clonal and true-to-type propagation of plants by a variety of tissue and cell culture methods is called as *in vitro* propagation, which is considered as the most efficient and commercial method of plant propagation. More precisely, it is called as micropropagation, which refers to the production of plants from very small plant parts, tissues or cells, grown aseptically in a test tube or containers under controlled nutritional, environmental and aseptic conditions. Tissue culture or *in vitro* culture is two broadly used terms for micropropagation, which basically include aseptic culture of various plant parts. Micropropagation was developed as the result of accidental findings of French Professor G. Morel in 1958, when he was attempting to obtain virus-free plants through meristem culture in dahlia and orchids. All the biological principles of micropropagation technique are based on the phenomenon of totipotency of cell, in when a cell of a plant has the capacity to regenerate in to a full-fledged plant having different organs. German plant physiologist, Haberlandt (1902) is known as the father of tissue culture technique, who for the first time coined the term *totipotency*.

Commercial status in micropropagation has been achieved in over 500 species and genotypes including herbaceous species foliage plants and fruit crops like banana, strawberry, papaya etc., and vegetables like potato, asparagus, etc. In fruit crops, different woody species, such as rootstocks of temperate fruit crops like apple, pear, peach and plum are also multiplied by micropropagation in some western countries. In India, there are over 130 commercial tissue laboratories, which mostly multiply fruit crops like banana, papaya, strawberry, sugarcane, medicinal and aromatic plants, spices, floricultural crops etc. At present, India produces about 80 millions plants per year; however, the installed capacity of such laboratories is 300 million plants per annum, which accounts to about twenty five to thirty percent of the installed capacity. About seventy five per cent share is from the private sectors and joint ventures.

Advantages and disadvantages of in vitro propagation

In vitro propagation offers some significant advantages in quality, quantity and economics over vegetative and seed propagation for many plant species. However, it has certain disadvantages also that can't be ignored if maximum potential of this technology has to be taken.

Advantages of Micropropagation

1. **Production of large propagating material in a shorter time:** Micropropagation has the advantage of producing a large number of planting material. Due to higher multiplication rate, millions of plants can rapidly be produced within a short time as compared to asexual and sexual methods of propagation.
2. **Production of disease free material:** Many diseases and the contamination of the source plant material are the limiting factors in conventional propagation methods. *In vitro* propagation techniques address these problems to minimize infection and produce disease free planting material in many horticultural crops.
3. **Production of true-to-type plant:** In micropropagation, there is gurantee of production of true-to-type propagation material having characters similar to mother plant. Now, methods have been developed and designed to provide the maximum number of true-to-type plants from a minimum number of selected mother plants.
4. **Ability to Air-Ship large quantity of planting material:** As many as 30.000-50,0000 micropropagated plants can be packed in to a cubic meter of shipping space. It has been found as the most efficient, quick and inexpensive method of transport of plants.
5. **Newly bred varieties/hybrids can be brought to markets quickly:** There is a great demand for the propagating material of the newly released varieties/hybrids, which can't be met with traditional methods of propagation. However, it is quite easy to meet this demand by exploiting *in vitro* propagation techniques.
6. **Plant production throughout the year:** In conventional propagation, seeds or other propagation material is available at a particular time of the year. However, under *in vitro* techniques, the process of multiplication continues throughout the year because of controlled environmental conditions.
7. **Useful in propagation of dioecious plants:** Under field conditions, male and female plants of dioecious species (date palm, papaya etc.) are to planted in a definite ratio and in general, male are required in a very less proportion, whereas these are produced in large amount under natural conditions. In micropropagation, only female or male plants can be produced in desired number by taking explant from the wanted ones.
8. **Production of secondary metabolites:** In tissue culture systems, many secondary metabolites are produced, which are used for the production of medicines by pharmaceutical industries.

9. **Breeding cycle is reduced:** Using tissue culture techniques, breeding cycle of the crops is reduced.
10. **Production of seedless hybrids:** By the use of embryo rescue technique, it is possible to develop seedless varieties as in grape.
11. **Useful in hard-to-propagate plant species:** *In vitro* propagation is highly useful for those plants, which are hard-to-propagate by vegetative methods or their propagation rate is very slow as in date palm, coconut etc.
12. **Production of homozygous plants:** By using protoplast culture, it is possible to raise homozygous plants easily, which takes years together through back crossing method under conventional breeding programme.
13. **To obtain rare hybrids:** In many interspecific and intergeneric crosses, fertilization occurs normally, but poor and abnormal development of endosperm causes pre-mature death of the hybrid embryos. The embryo culture technique is extensively used to rescue such hybrid embryos from wide crosses, which otherwise would abort.
14. **To overcome seed dormancy:** Dormancy in the seeds of deciduous plants can be reduced by growing excised embryos in the culture medium.
15. **Seed viability can be tested more rapidly:** The possibility of breaking seed dormancy by embro culture allows the use of this technique for testing the seed viability of a particular seed lot more rapidly.

Disadvatages of Micropropogation

1. **Contamination:** One of the greatest economic losses in commercial micropropagation is caused by endogenous and environment induced contamination of plant cultures. Contaminant once introduced in controlled environment, multiplies at a faster rate due to favourable environmental conditions. It is almost impossible to get rid of contaminant, if introduced in controlled conditions.
2. **Higher level of somatic variation:** In micropropagation, there are always chances of high level of variation in the plantlets. Variation may be due to excision of explant from diseased mother plant or incomplete monitring of production methods and plant material during micropropagation.
3. **Poor establishment of the plantlets in field:** One of the major concerns in raising material by micropropagation has been the poor establishment of the propagating material in the field.
4. **High production costs:** Expensive technology and facilities, and the labour intensive nature of tissue culture, offen render micropropagation economically unfeasible to many propagators.

5. **Poor response of woody perennials:** It is very difficult to raise most of the woody perennials through tissue culture techniques because of exudation of very high amount of phenolics from the explants, which hinder the establishment of explant in the culture media.
6. **Technical skill is required:** For raising plants through *in vitro* techniques, a technical expert is required, whereas for other methods of propagation, even illiterate peoples or farmers can raise plants.

Other implications of micropropagation

In addition to major use of tissue culture techniques for rapid clonal multiplication of plants, this technique is highly important for several other purposes, as under:

1. Production and maintenance of pathogen free stock plants.
2. Long-term *in vitro* conservation of germplasm.
3. Selection and regeneration of transgenic plants.
4. Conservation of pollen grains.

Principles of micropropagation

In raising plants through micropropagation, some fundamental principles are involved, which are discussed below;

1. Titipotency and plant regeneration

All *in viro* methods of plant propagation rely on the phenomenon of totipotency of plant cells. Totipotency may be defined as the capacity of plant cell, tissue or organ to have the capacity to regenerate in to a full plant. This term was given a German Plant Physiologist, Haberlandt in 1902, who is considered as the father of tissue culture systems. This can be achieved by excising the explant and placing it on a suitable culture medium.

2. Axillary bud proliferation

It has been observed by many biotechnologists that a very large number of axillary shoot buds grow extensively when shoot tips, or even smaller apical meristems are cultured in an appropriate culture medium, containing higher concentration of cytokinins. Both the excision of the explant and the cytokinin rich medium activate the bud, leading to proliferation of many side shoots, which can later be separated for further culture.

3. Organogenesis: Development of organs

Organogenesis from explants results in the formation of both shoot and roots. Formation of these organs takes place simultaneously, but either shoots or roots are formed first. However, in general, shoots are formed first than the roots.

4. Somatic embryogenesis

Somatic embryogenesis is different from organogenesis in that the regeneration and organization are bipolar i.e., the shoot and root meristems in the explants are formed simultaneously from proembryonic masses.

Basic requirements for micropropagation

For successful *in* vitro multiplication, the following pre-requisites are required:

i) A well equipped laboratory with all required equipment (balance, autoclave, laminar flow, culture trolley, glass wares, chemicals etc) and facilities (media preparation room, washing and drying room, inoculation room and culture room).

ii) Asepsis i.e. absence of any microbial contamination.

iii) A suitable culture medium, which depends upon plant requirement. The common culture medium, which are used commercially are MS (Murashige and Skoog) medium, B5 or Gamborg medium, Woody plant medium (WPM) etc.

iv) Controlled environment i.e., temperatures between 22° and 27°C, 60-80% humidity and light intensity of 20 to 200 μmol $m^{-2}s^{-1}$.

Stages of micropropagation

Micropropagation takes place in a series of morphogenetic processes, i.e. culture initiation to acclimatization of plants designated as Stage 0 to IV, which vary with plant species. The standardized technique is often referred to as a protocol in which the entire steps along with the media and cultural conditions are well defined and also reproducible when commercial production is practiced. Micropropagation takes place in the following stages;

1. **Stage 0:** This is a preparatory stage wherein the mother/stock plants are maintained under extreme hygienic conditions and prepared for establishing cultures.
2. **Stage I:** This stage is called as the culture initiation stage where any explant excised from the mother plant(s) is cultured *in vitro* to initiate healthy cultures (**Plate** 4).
3. **Stage II:** In this stage, shoot proliferation is achieved to produce multiple shoots, i.e., it is the most critical stage, which decides the commercial fate of any protocol. The number and rate of multiple shoot proliferation vary with species and variety.
4. **Stage III:** During this stage, *in vitro* rooting of the regenerated shoots is achieved to produce complete plantlets.
5. **Stage IV:** It is the last stage where the *in vitro* raised plants are subjected to acclimatization treatment so that they survive better in the field conditions.

Hardening or Stage-IV of micropropagation is considered as the most critical stage of micropropagation as it is the beginning of the biochemical and physiological processes in plants, which are necessary for survival and making them functional. During hardening, the *in vitro* raised plantlets are subjected to *ex vitro* conditions and thus they resort to develop functional root system, increased absorption of water and mineral nutrients, increased photosynthesis, etc.

The physiological and anatomical features of *in vitro* propagated plants necessitate the gradual weaning of such plantlets before transferring them to glasshouse conditions. This may include use of water permeable lids, bottom cooling, pre-conditioning with a higher level of sucrose for the development of well developed leaves and roots, photo-autotrophic micro-propagation by resorting to reduction in level of O_2 and increase in photosynthetic photon flux and CO_2 concentration, and use of growth retardants such as paclobutrazol, ancymidol, mepiquat chloride, diaminozide, ABA, etc. These growth retardants promote root formation, increase epicuticular wax and chlorophyll content of leaves and reduce the size of stomata. Besides, amino acids such as glycine-betaine improve photosynthetic activity and root elongation of such plantlets. Now-a-days, the emphasis is given on inoculation of microplants with beneficial microorganisms such as mycorrhiza, *Pseudomonas fluorescence, Bacillus* sp. etc. Beneficial interaction of microbes with the host plants result in enhanced survival, improved growth, and enhanced nutrient uptake in *in vitro* propagated plants.

After gradual weaning under *in vitro* conditions, plantlets are transferred to protected structures like glasshouse, polyhouse, humidity tents, wet tent cover, plastic covers over single flats, automated mist and high pressure fog systems. The high humidity of at least 85% is required for hardening- off at the initial stages. In addition, care has to be taken for avoidance of overheating of plantlets due to high level of solar radiation/irradiance (Plate 5).

Owing to huge requirement of planting material, Government of India through Department of Biotechnology has initiated the programme of establishing big sizes hardening facilities for the large-scale multiplication of forest trees and horticultural plants through tissue culture in the different agro-climatic regions. Six regional hardening units have been established at:

1. Tata Energy and Research Institute (TERI), Gual Pahari, Gurgaon, Haryana
2. Jai Narayan Vyas University, Jodhpur
3. Haryana Council of S&T, Hisar, Haryana
4. G. B. Pant Institute of Himalayan Environment and Development, Almora
5. West Bengal Council of Science & Technology, Kolkata
6. Regional Research Laboratory, Jammu (J&K)
7. Tata Energy and Research Institute (TERI), Guwahati.

Synthetic seeds

To avoid the cumbersome hardening processes, plant propagators are now shifting toward alternative options such as use of synthetic seeds or synseeds, which need no hardening and can be planted even directly under field conditions. Synseeds are defined as artificially encapsulated somatic embryos (**Plate 6**), shoot buds, cell aggregates or any other tissues that can be used for sowing as seeds and posses the ability to convert into a plant under *in vitro* or *ex vitro* conditions (**Figure 40**). The encapsulated seed also contains nutrients, growth regulators, pesticides, antibiotics etc., needed for the development of a complete plantlet from the enclosed somatic embryo (embryos developed from vegetative cells and tissues within *in vitro* system) or shoot bud.

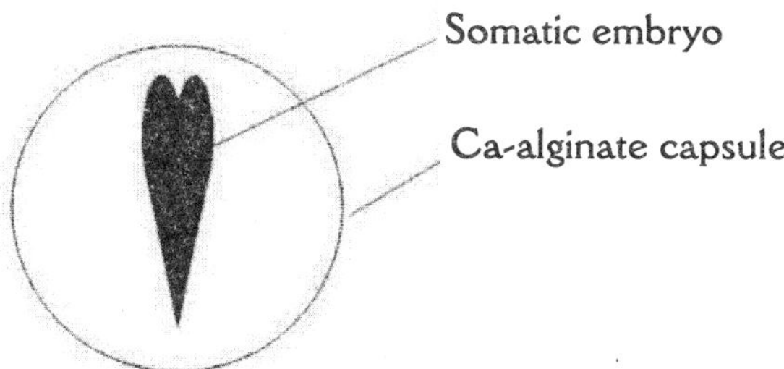

Fig. 40. Synthetic seed containing a somatic embryo.

Technique of synseed production

Shoot bud/ somatic embryos are allowed to disintegrate into individual seeds from the differentiated callus. These are then dropped in a solution of sodium alginate and added into the fresh medium along with somatic embryos/shoot buds. With the help of micropipette, individual somatic embryo/shoot bud is dropped into calcium chloride solution. The outer layer of each drop immediately forms biodegradable layer of calcium alginate. Later, these synseeds can be used for multiplication like any other sexually propagated seeds. This technique has been standardized in many horticultural crops including fruits, vegetables, ornamentals and aromatic and medicinal plants (**Table 8**).

Advantages of Synseed

- Combined advantages of clonal and seed propagation
- Introduction of agricultural chemicals and microorganisms
- Yield uniformity
- Advantages mainly with high-valued germplasm conservation and exchange
- Elimination of rooting and hardening phase

Table 8. Horticultural crops in which synseed technology has been standardized.

Plant	Propagules used
Fruits and plantation crops	
Apple	Shoot apices
Banana	Shoot apices
Tea	Somatic embryos
Coffee	Somatic embryos
Mulberry	Axillary buds
Mango	Somatic embryos
Vegetable crops	
Potato	Somatic embryos
Carrot	Somatic embryos
Tomato	Axillary buds
Brinjal	Somatic embryos
Ornamental crops	
Cymbidium pendulum	Protocorms
Cymbidium giganteum	Protocorms
Geranium	Somatic embryos
Lilium	Somatic embryos
Geodorum densiflorum	Protocorm like bodies
Spices, medicinal and aromatic plants	
Black pepper	Shoot buds
Cardamom	Somatic embryos
Cinnamon	Shoot buds
Basil	Nodal segments
Vanilla	Shoot buds

Major *in vitro* systems

Based on the type of explant, the different *in vitro* techniques have been employed in micropropagation of different plant species. A few of them are briefly discussed hereunder:

1) **Meristem tip culture:** This technique is widely used in horticultural plants like potato, dahlia, carnation and orchids. In this method, the meristem tip consisting of one or two pairs of leaf primordia are cultured in a medium. After a few weeks, the plantlets are re-generated and after hardening of the plantlets, these are transplanted in the soil under natural environmental conditions. Meristem tip culture is very useful for the elimination of viruses from infected plant material. Plants produced are free from pathogens and can be stored for longer period and in smaller space.

2) **Embryo culture:** Aseptic culture of embryo has been standardized in many plant species. The embryo is rescued while it is immature and before the degeneration is initiated. The excised embryo is cultured on the semi-solid or solid culture medium as the case may be. However, in embryo culture, the addition of some plant extracts like coconut milk or yeast extract in the culture medium enhances growth and development of the embryos. Embryo techniques are primarily useful in case of making distant crosses in hybrid production. It is also useful technique to overcome seed dormancy. In addition, several embryological problems can be studied through embryo culture. In grape, embryo rescue technique has been used world wide to produce seedless varieties by excising immature embryo in crosses between Seedless x Seedless, which is otherwise not feasible through conventional propagation or breeding techniques.

3) **Callus culture:** A callus is an unorganized mass of parenchymatous cells. In callus culture, the stimulation of cell division and multiplication is required and thus culture medium is accordingly supplemented with necessary growth regulators. For callus culture, auxins are required frequently followed by cytokinins. In callus culture, organogenesis occurs in two stages. In first stage, formation of meristems takes place and in the second stage, active growth of stem buds and roots takes place. As the nutritional requirement of these two stages is different, transferring of first stage to the second stage with culture medium having different supplementations (especially hormones) is a must.

4) **Anther culture:** Anther culture is used to produce haploid plants. Haploid plants are very useful in genetical and biochemical studies and for producing homozygous lines (within six months) of F_1 hybrids, which otherwise may take six generations or more with conventional methods of breeding. Guha and Maheshawari reported the first anther culture in Datura in 1959. In this technique, anthers are excised and cultured on to required and standardized culture medium. The stage, at which the anther is excised, depends on different species. However, in most cases, it is excised at or near microscopic mitosis, when the pollen grain passes from uninucleate to the binucleate condition. As a result of cell division, multicellular masses are developed, which may develop in to embroids or callus. Plantlets are developed from callus by organogenesis. Haploid plants are treated with colchicine for chromosome doubling and for producing the true breeding diploid plants.

5) **Cell culture:** Cell suspension cultures are obtained by homogenizing a piece of callus into liquid medium and shaking with shaker until medium becomes cloudy with suspended cells. Due to recent advances in Plant Biotechnology, it is now possible to derive full plant from a single cell. This technique is primarily useful to plant breeders because now it is possible to induce desirable variability in individual cell, which can be regenerated into a full plant.

6) **Protoplast culture:** In protoplast culture, the rigid cell walls are first removed either mechanically or enzymatically so as to expose the protoplast. The exposed protoplast is then cultured onto a suitable culture medium. Individual protoplast of the plants has been regenerated into full plant as in *Nicotiana, Brassica* and aparagus etc. In plant breeding systems, protoplast culture has many advantages i). A large homogenous population (10^6 to 10^7) of plants can be obtained from a small sample (1 g leaf), ii). Fusion of two protoplasts of two different plant species is possible, which may bring a greater variability in plants.

7) **Ovule culture:** In this method, unfertilized ovules are excised and cultured into medium. This technique has potential application to produce hybrid seeds in wide genetic crosses, which is rather difficult through conventional means. It is also efficient method to obtain plants in self-incompatible species. The technique of growing ovules in sterile culture is not a difficult job. However, one should have sound knowledge of the flowering and fertilization timing for isolation of the ovule at a proper time.

Status of micropropagation industry in India

At present, an estimated 80 millions tissue cultured plants, from 130 tissue culture companies, are produced every year in the country. The major share of produced plants is occupied by banana (41%) followed by sugarcane (31%), ornamental plants (14%), spices (6%) and medicinal plants (4%). To bridge the gap between research and field testing, and to check the clonal fidelity in such planting material, the Department of Technology, Government of India, has set up pilot-scale facilities including Micropropagation Technology Parks (MTPs) at the The Tata Energy and Research Institute, New Delhi; National Chemical Laboratory, Pune, and National Facility for Virus Diagnostics and Quality Control of Tissue Culture Raised Plants (NFVDQCTP) at IARI, New Delhi, with 5 satellite centers NCL, Pune; TERI, New Delhi; IHBT, Palampur; IIHR, Bangalore and SPIC, Chennai.

The MTPs are established for mass production of elite planting material employing tissue culture technique and thereby helping in transfer of technology from laboratory to land, while centers under NFVDQTP work on testing clonal fidelity using standard molecular markers techniques for finger printing such as RAPD, AFLP and SSR and developing diagnostic kits and standardizing quality testing parameters for identified plants.

Among fruit crops, strawberry was the first fruit crop to be propagated commercially through micropropagation. Micropropagation techniques have been standardized for many temperate, tropical and sub-tropical fruit crops. In India, tissue culture technique has also been perfected in banana, grape and papaya. Shoot tips and two nodal micro cuttings are highly suitable explants for faster and disease free production of grape plants through tissue culture. Salt tolerant rootstock of grape (Salt Creek and St. George) has great demand and *in vitro* propagation has been successfully used. Shoot tip culture technique has been demonstrated in papaya to produce female and male plants in the desired ratio. Similarly, pineapple is now being commercially propagated through micropropagation. In hard-to-root fruit species, like date palm and coconut, *in vitro* techniques have been perfected only recently. Scientists in India and different parts of the world have been trying hard for the micropropagation of mango and perennial forestry plants through tissue culture but the success has been limited due to recalcitrance and browning of explants.

* * *

12 Nursery Establishment and Management

Establishment of Nursery

It is now well recognized that we should increase area under horticultural crops, in order to step up the production of fruits, vegetables, flowers, plantation crops, spices and condiments and medicinal plants. However, the greatest bottleneck in the expansion of area under these crops is the non-availability of true-to-type plants in adequate number. For which, the importance of nursery has been realized. A nursery is a place, where seedling, saplings, trees, shrubs and other plant materials are grown and maintained until they are placed in a permanent place. Unlike most of the agricultural crops, establishment of an orchard is a long-term investment and any mistake made initially can't be rectified in the later stages. Hence, the availability of a quality planting material is the foundation on which an orchard is built. Unfortunately, most of the present day orchards in our country have been established from inferior stock. Thus, the fore-most pre-requisite to assure success of horticulture industry mainly depends on pedigree plants. Nursery is a place where one can produce true-to-type plants by gaining technical skill, maintaining plants properly (Plate 7). Like establishment of orchards, setting up of a nursery is a long-term venture and needs lot of planning and expertise. Mistakes committed at an early stage on any aspect, like selection of soil, raising of right kind and varieties, plant protection measures, etc., will adversely affect the returns from the investment, besides wastage of time and energy. Thus, one should give utmost attention and care when nursery is to be established.

Classification of the nursery

On the basis of size and purpose and nature of business, the nursery can be classified as follows:

A. Classification based on size and purpose of nursery

(i) Home nursery

(ii) Commercial nursery

B. Classification based on nature of business.

(i) Wholesale nursery

(ii) Retail nursery

(iii) Landscape nursery

(iv) Agency nursery

(v) Mail order nursery

C. Classification based on size and purpose

(i) Nursery for home purpose: Small area in the garden in which the plants are grown to fulfill the of grower's own demands. In this type of nursery, usually costly methods of nursery practices are followed for various types of planting material of high quality to meet his own demand.

(ii) Nursery for commercial purpose: Such nurseries are meant for earning monetary profit on the investment. Such nurseries have a large area and usually costly nursery practices are avoided as a result the control on quality of planting material is also reduced. Commercial nursery may be urban locality based (urban nursery) or rural based (rural nursery).

a. Rural nurseries: Such nurseries are generally located in a village near some high way or near a railway station. Usually, the size of rural nursery is large because land and labour is not a problem in rural areas. The planting material in such nurseries is sold out at cheaper rates because the cost of raising the planting material is less.

b. Urban nurseries: Nurseries located in a town or cities are called as urban nurseries. The size of such nurseries is usually small because the land is costly, and is not available easily in the urban localities. Due to higher production cost, the planting materials sold are at higher rates.

D. Classification based on the business

(i) Agency nursery: The agency nursery sells its stock through agents or sales representatives. Such nurseries are highly specialized and are usually few in numbers.

(ii) Landscape nursery: The landscape nursery should be located near a populous town or city, because urban people require the landscape plants for beautifying their houses. Such nurseries should be located on the outskirts or a city or in a town as it facilitates the clientele to obtain landscape service.

(iii) Mail order nursery: It is a specialized wholesale nursery. It depends primarily on a catalog display of the stock it offers for sale. Customers order from the catalog and receive the plants through mail or parcel service. These nurseries are also located in a locality, where land is relatively inexpensive and sufficient labour, water and transportation facilities are available.

(iv) **Retail nursery**: The retail nurserymen purchase plants from the wholesale nursery and sell them to the consumers. Therefore, these nurseries are largely dependent on wholesale nurseries for their trade. Retail nurseries must be located near a town or a city. These nurseries also keep goods like fertilizers, seeds and tools etc. required for raising garden plants.

(v) **Wholesale nursery**: In this type of nursery, the plants are produced in large quantity for sale to the retailer. Wholesale nurseries are usually located in rural areas, where the land and labour are available at cheaper rates.

Pre-requisites for nursery establishment

Following pre-requisites are considered important for nursery establishment:

(i) The locality where fruit growing is being practiced already provides an ideal site for setting up a nursery.

(ii) As far as possible, the nursery should be located in the important production areas.

(iii) The site should be well connected by different means of communication and must be easily accessible.

(iv) Climatic conditions should be favourable for the plants to be propagated.

(v) The nursery should have its own resources to provide parent material (mother plants) for propagation (**Figure 41**).

(vi) The locality should have adequate supply of sweet water.

(vii) The soil should be deep, fertile, well drained and free from soil borne pathogens.

(viii) There should be sufficient provision for different propagation structures like green houses, glass house, lath house etc.

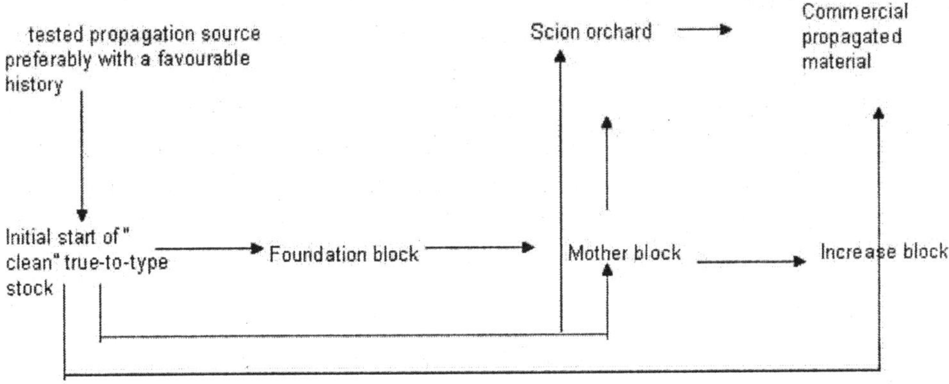

Fig. 41. Method for nursery production.

(ix) Sufficient labour, budder and grafters should be available to handle different operations.

(x) The materials like fertilizers, pesticides, growth regulators, grafting wax, lanolin paste etc. and equipments etc. should be readily available in the locality.

Importance of nursery

1) Nursery is the only place, where people can get genuine planting material of their choice.
2) The young plants can be easily maintained in the nursery. Similarly, it is easy to look after the young and tender seedlings in the nursery rather than in the field.
3) Propagation of plants by asexual means require special skill and care of plants before transplanting in the field, which can be easily done in nursery but not at consumer's level.
4) Seasoning of the seedlings against natural calamities is only possible in the nursery.
5) Many plants do not respond to direct sowing e.g. cabbage, tomato, papaya and conifers, in the field as compared to transplanting of seedling raised in the nursery.
6) Cuttings of different horticultural and forestry plants for rooting are first planted in nursery for better care and management.
7) For hardening of seedlings/grafts/layers, nursery is a pre-treatment place.

Parts of a nursery

A modern nursery should have the following parts:

1. Permanent structures

An ideal nursery should have the office, sale counter, packing shed, store house, implement shed, bullock shed and residential quarters etc. The office is the most important amongst all the building structures because all the business transactions are to be made from the office. It should look beautiful and should have the ability to attract customers. The size of office depends on the size of the nursery and should be located at such a place from where the condition of nursery and work can be supervised properly. However, both the store and office can be in the same building. Store can be utilized for keeping packing material such as baskets, labels, nails, ropes, equipments, fertilizers and planting material etc. The owner of the nursery can use the office. Further, office should be near to the gate of the nursery so that visitors have no difficulty in locating it.

2. Progeny block

A nursery should have progeny trees of all the promising plants and varieties that can be grown in that particular area. In progeny block, true-to-type mother plants are maintained in the nursery. According to the choice of customers, suitable plant types with existing superior varieties should be collected and maintained in the progeny block. The plants in the progeny block should be healthy, disease free, genetically pure and free from the attack of insect-pests and diseases. Proper care regarding control of insect-pests and diseases should be taken in respect to mother plants in the progeny block. It is emphasized that small area should be given to each type of fruit with different variety. The plants of each variety should be labeled and maintained correctly.

3. Propagation structures

Nursery is a place where raising of plants is carried out throughout year. However, in few months of the year, open conditions are not suitable to raise plants successfully. The hot desiccating winds of May-June and cold freezing waves of December-January may harm the plants and their subsequent growth. Hence, there should be the provision of modern propagation structures, like green house, hotbed, cold frame, lathe house, net house and mist chamber etc. These structures provide optimum controlled conditions (light, temperature, moisture and humidity) for seed germination, rooting of cuttings and for hardening of young seedlings before transplanting them in the field.

4. Seed beds

Seeds beds can be accommodated in a comparatively smaller area. They should be nearer to a source of water supply and to the office so that they can be kept under proper vigil. The beds should be raised enough to avoid water stagnation due to rain and excess watering. Seedbeds should be located in an open situation for better germination of seeds and to avoid infestation by 'damping off' disease.

5. Nursery beds

Seedlings are lifted from the seedbeds and transplanted in the nursery beds. Nursery beds should be slightly raised, with 3 m length and 1 meter width. It is better if the nursery beds are divided in to sections. For example, one section can be a mango seedbed area; a second section can be set apart for rooting of cuttings, another for fruit plants and still others for citrus plants etc. This order should be changed next season to have a rotation and thereby keeping the fertility of the soil intact. A nursery bed should be prepared by repeated ploughing and adding appropriate amount of organic manures and fertilizers. There must be adequate facility of proper drainage in the nursery beds. The nursery beds should be laid out in such a way that there is an access to all the beds through roads or paths so that workmen may not have to walk over bunds or in water channels. Nursery beds should be located in an open area near to a water source.

6. Pot yard

The pot yard generally used for the tender plants, which require shade as compared to hardy plants. Therefore, pot yards should be in shade and near to water source for frequent watering. Trenches can be provided for keeping the potted plants closely packed together.

7. Packing yard and working shed

The packing yard is used for packing the plants before sale or dispatch to out stations. The yard can be combined with the working shed. In packing yard, there should be plenty of space to enable a number of workers for sorting out and packing the plants with ease.

8. Water source

The timely availability of good quality water in sufficient amount is an integral part of nursery production. In nursery, several plants of different stages of growth are raised, therefore, timely supply of quality water is an essential. Thus, proper arrangements should be made to meet the water requirement of a nursery. For this purpose, provisions for water storage tank, motor or diesel engine operated tubewell should be made. The water channels should also be evenly distributed. These channels should be connected to every part of the nursery for timely use.

9. Compost pit

Nursery production of horticultural and forestry plants require huge amount of organic manures, like F.Y.M., compost, leaf mould etc. for different purposes. Therefore, arrangements should be made at nursery level to produce enough quantity of compost etc., for own purpose. One compost pit of permissible size should be located at any corner of nursery layout. At the same time, several waste products of nursery can also be utilized for the same and extra effort to dispose off waste material is not required.

A plan for ideal fruit plant nursery has been given in figure 42.

Factors affecting the nursery establishment

Following factors have been found to affect the nursery establishment,

1. **Site and location:** Major factor, which have an important bearing on the success of a nursery in a particular locality is that it should be located near a city or town connected with metalled road, or near a railway station so that it is easily accessible and the customers can visit it conveniently. The area selected for nursery establishment should be reputed one or other kind of business. It should be in known locality. The availability of skilled labour and other inputs should be assured.
2. **Soil and sub-soil:** For establishment of a nursery, both soil and sub-soil should be suitable and the pH must be in an optimum range. The soil should be physically

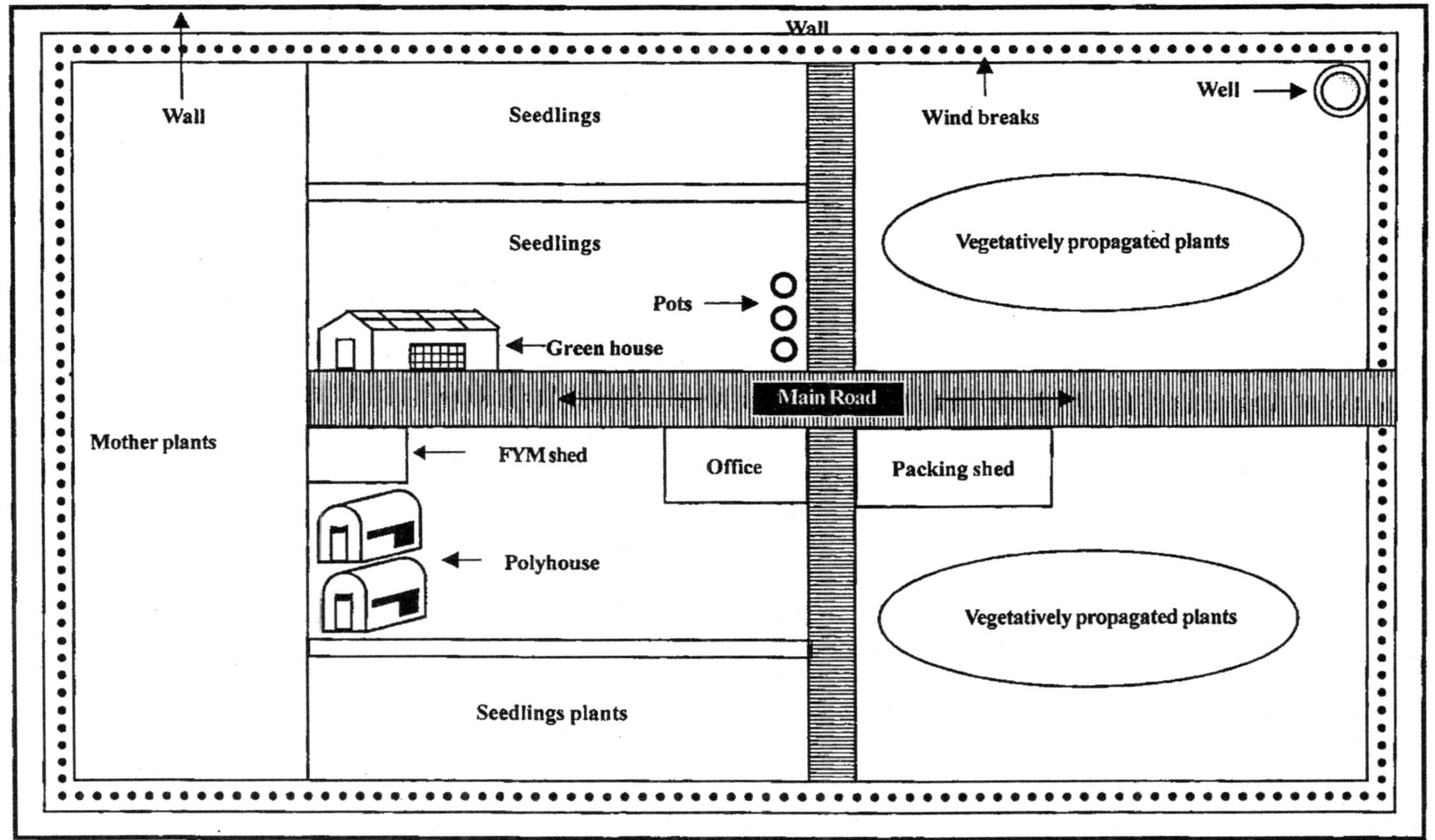

Fig. 42. Layout plan of an ideal fruit plant nursery

good and should have proper drainage. Usually, loamy soils of high fertility with pH range of 5.5-6.5 are preferred. A clay loam soil is advantageous, as in this type of soil, the lifting of evergreen plants with earth ball intact is easy. However, soils with hard and compact sub-soil layers should be avoided. Soils with impregmented layers of calcium carbonate may be avoided as it affects permeability and aeration in the soil, resulting in poor root development. Highly acidic, alkaline and saline soils are not preferred for nursery purpose.

3. **Demand of plants and varieties:** While making choice of kind of plants and their varieties, one has to take into consideration a number of factors. Demand for fruit plants of different kinds and varieties is the most important factor. In the nursery, only those plants and varieties should be produced in bulk, which are in high demand in that particular locality.
4. **Irrigation arrangements:** Nursery stock and young plants need frequent irrigation to make them grow well. Therefore, provision of satisfactory irrigation arrangements is essential for the success of a nursery. In canal irrigated areas, where canal closures occur and in those areas where there is paucity of irrigation water, it is advisable to construct a water tank of adequate capacity, so that water stored in tank could be utilized at a time of water scarcity. In place where, tube wells are operated with electric motors, there should be stand by arrangements of diesel engine, so that the same could be utilized during electric shut-offs.
5. **Availability of chemicals, manures and fertilizers:** In the nursery, many chemicals, manures, fertilizers and plant protection are required for timely application and management of seedlings and plants in a sound manner. Organic manures like farm yard manure, leaf mould, compost, and fertilizers like urea, single super phosphate, potassium and several fungicides, insecticide, weedicides are required and these must be timely available in the locality in sufficient quantities.
6. **Trained gardeners and skilled labourers:** The orchardists who want to take up nursery production work, should be well conversant with different propagation techniques and other nursery operations. They should have with them trained budder/ grafter, who could perform these operations with good success.

Nursery management

Genuine and quality nursery plants are the foundation of successful fruit industry as a business, which can be achieved by producing large number of reliable, vigorous, disease and insect free, and true-to-type plants in shortest possible duration. To raise plants in the nursery seems to be easy, but to maintain them in good state is very tedious. A seedling/ sapling is considered of high quality if it meets the expectations or standards of performance on a particular site. Raising of seedlings and maintaining the young plants in a nursery is a continuous process. Therefore, one should have complete idea about sowing time, sowing depth of the seed etc. to raise the seedlings effectively. Similarly, knowledge about transplanting of seedlings, their handling and further management in the nursery is equally important for raising quality material.

Sowing of seeds in the nursery

In sexually propagated plants, the seeds are first sown in the seedbeds and after attainment of sufficient growth of the seedlings, they are transplanted in the permanent site. Similarly, rootstocks of fruit plants are first raised in the nursery and at optimum growth of seedlings, they are subjected to budding, grafting etc., with suitable and desired scion variety. Thus, it is considered as a specialized job because one has to acquire fair knowledge about the plants to be raised and their cultural requirements in the nursery.

A. Preparation of Seedbeds

The most important steps for raising seedlings in the nursery are to select a proper site for seedbeds, seed sowing, care of seedbeds, care of seedlings and then transplanting of the seedlings at aright place and at right time. Seedbeds should be made in a locality where the soil is fertile, free of clods, and rich in nutrients and has adequate drainage and aeration. The beds should be able to provide good growth to the seedlings. For preparation of beds, the soil should be prepared thoroughly by repeated ploughings. During land preparation, suitable chemicals should be added to protect the seeds and seedlings from the attack of insect-pests, diseases and weeds. The soil should be leveled and the beds should be of desired size and sufficiently spaced in order to facilitate nurserymen to care all the seedlings from either sides of the bed. Usually, the beds of 10' x 3' x 6 " are preferred for raising vegetables and flowers seedlings.

B. Seed sowing

The seeds are sown either in seed bed, polythene bags or *in situ*. Seeds of tropical and sub-tropical fruits are sown during monsoon (June-July) or in the beginning of spring (Feb-March). Generally mango, kagzi lime and jackfruit seeds are sown during June-July while that of guava, *ber* and aonla during February-March. Seeds of temperate fruits are generally available during June to October and their sowing should be done after the dormancy period is over (Feb.-March). Seeds of some fruits (citrus, mango, loquat, litchi, jackfruit, etc.) lose viability quickly, so these should be sown immediately after their extraction from the fruit. Under north Indian conditions, the seeds of different citrus types are available during winter months. Germination of seeds is low due to the prevailing low temperature. Hence, use of alkathene tent on seed-beds during December-January has been found useful in increasing germination percentage and faster growth of the seedlings.

The seeds are usually sown at a depth 3-4 times of its size. Sowing may be a little deeper in light soils than in heavy soils. It is essential to maintain proper plant density in bed as crowding may result in weak and lanky seedlings. Watering, shading, or sun exposure, incidence of diseases and frost affect seed germination and subsequent seedling growth. Rootstock seedlings are usually kept in seedbed for 1-2 years. In fruits like walnut, pecanut, jackfruit and *ber*, sowing of seeds is done *in situ*. In rocky soils of Orissa and Gujarat, grafting is done on *in situ* sown mango seedlings.

After germination, the seedlings must be gradually exposed to light, temperature. Over watering should be avoided. The frequency and amount of irrigation should be increased with the growth and development of seedlings. Protection of seedling from frost during winter and from hot desiccating winds during summer is necessary. Similarly, the seedlings at younger stage may be infested by some insect-pests and diseases (especially damping off) and thus spray of seedlings with insecticides and fungicides should be taken on priority.

C. Hardening of seedlings

The seedlings raised in the nursery are very succulent and delicate and they can not withstand shock of transplanting unless they are properly hardened before transfer in the permanent site. It can be achieved by withholding water supply and reducing temperature or by combination of both. Thus, the reduced growth rate enables the plants to withstand desiccation and helps in the proper development of roots. However, over hardening may be avoided at every cost as it may result in extremely stunted plants. Thus, hardening of the seedlings should be done gradually for streamlined growth reduction.

The practice of hardening results in

1) Reduction in growth rate of the seedlings
2) Thickening of the cuticle
3) Increase in waxy covering on the leaves
4) Increase in the dry matter content of the seedlings
5) Decrease in transpiration rate
6) Increase in the levels of sugars in the leaves

All these conditions are essential to promote the seedlings to withstand the shock of transplanting, which would help in their better establishment in the field.

D. Transplanting in field

Plants raised in beds are finally transplanted in the permanent locations i.e. in the field. Sometimes seedlings are transplanted once or twice before setting in the field. Usually seeds are sown thickly in the seedbeds and as soon as the seedlings become crowded, they are transplanted to another bed or pot. This practice in common in flowering annuals plants and vegetables and is known as *pricking*. When the seedlings attain good size, these are transplanted in the field. During transplanting, maximum possible soil should be retained with the seedlings. After transplanting, plants are firmly pressed and watered immediately. To avoid wilting after transplanting, a provision of shade should be made, which should be removed towards the evening. The transplants should be irrigated frequently. However, water stagnation should be avoided at every cost. A starter or booster solution containing mineral nutrients can also be given. Usually, there is high rate of mortality of seedlings after transplanting. The major reasons for failure of seedlings in transplanting are:

1. If the plants do not have well-developed roots.
2. If the seedlings are not lifted with optimum earth ball before transplanting.
3. If the roots of the seedlings are soft and succulent.
4. If the plants are not watered adequately.
5. If the plants are not pressed firmly to avoid air space
6. If the plants are not given proper shade initially.
7. If the plants are not protected from scorching sun and wind.
8. If the environmental conditions existing in nursery bed are different to those in the field.
9. If the seedlings are not watered immediately after transplanting.

Lifting of nursery material

The plants raised in the nursery are to be lifted for their permanent place of establishment or for sale. The careless lifting may affect the establishment of nursery plants adversely. To achieve high rate of success in the field, it is always advisable to lift the saplings with a ball of earth. It is particularly true for evergreen plants like mango, guava, citrus and litchi etc. However, the deciduous plants like peach, grape, apple, kiwifruit, almond, walnut etc., can be lifted bare rooted. To avoid the damage to the roots, it is advisable to irrigate nursery, 3-4 days before the actual date of lifting the plants. The lifting of sapling should however be avoided in case there is heavy rain because earth ball does not form properly if soil is too wet. One of the oldest suggestions is to cut the tap root system of the evergreen plants, especially mango atleast one month before the actual date of lifting. This practice helps the sapling to tolerate the shock of lifting. Moreover, there is development of secondary roots in the sapling, which helps in better survival rate in the field.

After lifting, the seedling/sapling should be properly packed. The bare rooted saplings (grape, peach, apple, pear etc.) should be packed in moist sphagnum moss grass and wrapped with hessain cloth to avoid drying of the roots. The sapling of evergreen plants along with earth ball should be wrapped with moist grass or rice straw before putting in to polyethylene bags and then kept in basket. The strawberry runners are first packed in moistened moss grass and kept in hessain cloth bags or in baskets for long distance transportation.

Storage of saplings

Dry air and high temperature causes desiccation of the plants lifted from the nursery. Before sending them to distant places, these should be stored in a storehouse. The storehouse should have proper ventilation and it should be maintained at low temperature (2-3°C) and high relative humidity (85 to 90 per cent). These environmental conditions reduce the transpiration and respiration rate, avoids withering or desiccation of the seedlings/ saplings, thereby enhances their survival rate in the field.

Packing of nursery material

Packing of nursery material has to be done from time-to-time. Usually, much emphasis should be given while sending plants to a distant place but it should not mean that plant to be sent to nearby places should no be packed properly. To have a better price of the products, a nurseryman should pay high attention to the packing of nursery material. A packing material should have the following characteristics:

- It should be cheap and easily available.
- It should be highly suitable for the material to be packed.
- It should deliver the plants conveniently without affecting the quality.
- It should look attractive to the indentors.
- It should ensure protection against drying out and mechanical injury.
- It should be well adapted for convenient and economical handling.
- It should be well adapted to transport, loading with security and economy at a space.
- It should not make the material bulky rather it should add minimum weight to the material.

Packing either of seeds or plants or bulbs etc., should be done carefully so as to supply the planting material in safe and sound condition. It is particularly true when planting material has to be sent to distant places. For distant places, majority of material is sent by rail and trucks, where it is handled very roughly. And therefore proper packing becomes more pertinent

Packing materials

The different packing materials are used for packing the nursery plants are hessain cloth, sacking cloth, plastics, paddy straw, dried grass etc. Hessain and sacking cloth is traditionally used by almost all the nurserymen. It is used in balling and burolopping of the plants. The hessain cloth is made from the best jute fibres, whereas sacking cloth is made from the raw grade jute fibre. Plastics and its derivatives are used for packing nursery material by the modern nurserymen. Different plastic materials like low-density polyethylene (LDPE), high-density polyethylene (HDPE), polypropylene (PP) and Nylons etc., are used for packing purpose. Paddy and wheat straw and hay, dry grasses are used for wrapping the earthen ball of the saplings. Similarly, moistened moss grass is used to wrap up the delicate planting material like vegetable and ornamental flower seedlings and strawberry runners before packing them in hessain cloth.

* * *

13 Media and Mixes for Propagation

Nurserymen raise new plants in their nurseries by different means of propagation. Hence, they use different kinds of media and mixes for the propagation for growing these propagated plants. However, one medium may be suitable for seed sowing and the other for cuttage or layered or other propagules. Similarly, plants of one species may do well in a particular medium but may be a totally failure for the other. Hence, the choice of the use of a medium depends on the plant species, method of propagation and availability of a medium. However, an ideal medium should meet the following criteria;

1) The selected medium should be sufficiently firm and dense so that it holds the seeds or other propagating material (cuttings, layered plants, stools etc.) firmly in place during the process of seed germination or rooting.
2) The volume of the medium should remain constant upon drying or wetting.
3) It should have enough ability to retain and supply ample amount of moisture to the seeds or other propagating materials.
4) It should be porous enough to permit adequate aeration and to drain out excessive moisture.
5) The medium must be free from seeds of weeds, nematodes and pathogens, which affect the germination of seeds or rooting of the propagules adversely.
6) Its pH should range from 5.5 to 6.5.
7) It should be of consistent quality from batch-to-batch and it should be reproducible.
8) The selected medium must not have higher concentrations of salts.
9) It should be highly decomposed, preferably with 20:1 C:N ratio to prevent N immobilization and excessive shrinkage.
10) It should have high cation exchange capacity so that it retains the mineral nutrient.
11) It should be easily available.
12) It should not be costly.

13) It should be able to provide adequate nutrition to the germinating seeds or cuttings.

14) It must be capable of being pasteurized with steam or chemicals without harmful effects.

15) It must be easy to wet and should have the ability to retain sufficient moisture to reduce frequent watering.

Criteria for the selecting a propagation medium

Before using any propagating medium, following criteria should be kept in mind, while selecting it:

1. Always select a locally available material for growing plants or raising seedlings.
2. Before buying, check the quality of the medium.
3. Always take guidance of a technical expert or a person dealing in nursery business.
4. The selected medium should have sufficient porosity and ideal pH (5.5-6.5).
5. The selected medium should easily mix with other growing media.
6. It should meet all the requirements of the seeds to be raised or crop to be propagated.

Commonly used propagation media

Propagation media used in raising horticultural and forestry plants in the nurseries mainly consist of a mixtures of organic and inorganic components. The organic components include peat, bark or sphagnum moss. Sometimes saw dust, leaf mulch, paddy straw, wheat straw, and paddy hulls are also used but because these decompose readily and compact easily, which decreases pore space and aeration. For this, some coarse mineral component should be used for increasing aeration and to improve drainage. Such mineral components are sand, grit, pumice, perlite, vermiculite and rock wool etc.

Selection of medium is affected by a number of factors, like species to be propagated, type of propagule, season of propagation, propagation system, cost and availability of the medium. However, the following media can be used in propagation of various horticultural and forestry plants;

Soil

The portion of earth's crust, which is formed by the decomposition of rocks and minerals by physical, chemical and biotic forces, is called as *soil.* The soil is mainly composed of solid, liquid and gaseous states. For proper growth of the plants, all these contents should be in a proper proportion. Solid portion of soil has both organic and inorganic matters, which determines the structure and texture of the soil. The inorganic part is mostly the residue, which is the resultant of decomposition of parent rocks by way of weathering process. The organic portion of the soil constitutes of both living and dead organisms.

Fungi, bacteria, worms, viruses and insects generally constitute the living organic matter, whereas, the dead organisms in various stages of decay make up the *humus.* The liquid part of the soil is mainly water, which contains dissolved mineral elements and dissolved O_2 and CO_2 gases. The gaseous portion of soil is very important for plant growth as in water logged (poorly drained) soil, water replaces air, thus plant roots are deprived of essential microorganism, which are otherwise required for proper growth and development of the plant.

Soil texture refers to the proportion of individual soil particles, which is an important physical property of the soil and determines the success or failure of seed or cutting. A soil having 40 per cent sand, 40 per cent silt and 20 per cent clay is considered best for seed germination of many plant species. Sandy loam soils are excellent ingredients for preparation of soil mixtures for container growing plants. The pH of soil also determines the success or failure in seed germination or rooting of cuttings. A soil having neutral pH (5.5 to 6.5) is always preferred.

In addition to texture, the structure of soil, which refers to the arrangement of soil particles in a soil mass, has great bearing on the germination of seed and rooting of cuttings. Water holding capacity of soil is usually determined by its structure. Thus, a soil having good texture should be preferred for using as a propagating medium. In general, heavy soils should be avoided, as working with these soils is very difficult. Moreover, these soils form heavy clouds and have very high water holding capacity, which is injurious, both for seed germination and rooting of cuttings.

Sand

Sand is formed as a result of weathering of rocks and its composition depends upon the type of the rock from which it has formed. The size of sand varies from 0.05 to 2.0 mm in diameter. Quartz sand, which chiefly consists of silica compounds, is mainly used for propagation by the nurserymen. Sand is the heaviest of all the rooting media and therefore it should be used in combination with some other organic material.

Vermiculite

Vermiculite is a micaceous mineral, which is widely used as propagating medium by the nurserymen in many advance countries like the USA, Japan, Germany etc. Chemically, it is a hydrated magnesium-aluminum iron silicate, and it expands markedly upon heating. It is very light in weight, neutral in reaction and insoluble in water but absorbs very high quantity of water. It has a high cation exchange capacity and can hold nutrients for longer time and releases them later. It contains sufficient quantity of Mg and K for supply to plants. Horticultural vermiculite has four grades depending upon the size of the particles. NO. 1 grade has particles from 5-8 mm in diameter, NO. 2 from 2-3 mm, NO. 3 from 1-2 mm and NO. 4 from 0.75-1 mm diameter. In general, No. 2 grade vermiculite is a regular horticultural grade and is widely used, but No. 4 is the most useful as a seed-germinating medium. However, nurserymen are advised not to use non-horticultural grade vermiculite as it is treated with chemicals, which are toxic to the plants.

Peat

Peat contains mainly the partially decomposed parts of aquatic, marsh, bog, or swamp vegetation. Its composition depends upon the vegetation from which it has originated, state of decomposition, mineral content and degree of acidity. Peat is mainly used as a propagation medium in the USA and there it has been classified in to different types *viz. moss peat, reed sedge peat* and *peat humus.*

Moss peat, which is commonly called as *peat moss,* is the least decomposed peat, which is mainly derived from sphagnum or hypnum or other mosses. It is highly acidic (pH 3.2-4.5) but has high moisture holding capacity. It contains small amount of nitrogen but little or no P and K. This type of peat moss is the most commonly used in horticulture and forestry nurseries, the coarse grade being the best. When peat moss is to be used in mixes as a propagtion medium, it must be broken down in to pieces and moistened before use.

Reed sedge peat is mainly derived from the remains of grasses, reeds, sedges and other swamp plants. It varies considerably in colour and composition, ranging from reddish brown to blackish colour. Its pH ranges from 4.0 to 7.5 and has high water-holding capacity. It is rarely used for propagation purpose.

Peat humus is the highly decomposed material either of hypnum moss or reed sedge peat, which has brown to black colour and has very poor water holding capacity.

Peat is not a uniform product and can be a source of weed seeds, insect-pests and pathogens. Thus, it must be pasteurized before its use as a propagation medium. Moreover, it is quite expensive and hence it should be used in combination with other media or mixes.

Among different types of peats, moss peat is very popular among nurserymen of most of advance countries because of the following reasons:

1. It is an uniform product.
2. Mostly, it is free from insect-pests, diseases and weed seeds.
3. It can be easily mixed either with sand, saw-dust, perlite, vermiculite or other propagating media.

Sphagnum moss

The sphagnum moss is commonly found in hilly tracts of India. The dehydrated moss is either residue or living part of acid bog plants of genus Sphagnum, like *Sphagnum papillosum, S. capillaceum* and *S. palustre.* Sphagnum moss is a relatively sterile, light in weight, having very high moisture holding capacity. It has a pH of about 3.5 to 4.0. It is widely used for keeping the live material moist for distant transport and in air layering (goottee) of plants. It can also be used as a rooting medium for cuttings. Moss should be shredded either by hand or mechanically before use. Similarly, it should be moistened with water before use. Sometimes, moss contains a strain of Streptomyces bacteria, which inhibits 'damping off' of the seedlings in the nursery.

Pumice

It is most widely used alone or in combination with peat moss for seed propagation in the USA and some other advance countries. It is of volcanic origin and mined in to commercial grade, which chiefly contains silicon oxide and aluminum oxide with varying amount of iron, calcium, magnesium, and sodium in oxide form. Pumice is graded into different grades but is not heat-treated. It increases aeration and drainage in rooting mixtures if used in combination with other propagating media.

Perlite

Perlite is a gray-white silicaceous material of volcanic origin, which is mined from lava flows. It is prepared by heating glassy volcanic lava in furnace at 760°C. Perlite is a sterile medium with pH of 6-8, high cation exchange capacity, water holding capacity but with no buffering capacity. It contains no mineral nutrient and is highly useful in increasing the aeration in a mixture, when used in combination with other propagating media. Perlite is available in different grades with particle size ranging from 3-8 mm but usually a particle size of 1.6 to 3.0 mm in diameter is mostly used by the nurserymen. It is considered as an ideal rooting medium for rooting of cuttings if used in combination with peat moss.

Leaf mould

Leaf mould is prepared by decomposing fallen leaves of the plants. Sometimes, chemical fertilizers, farmyard manure (FYM) or cow dung are also added to it. Its colour and composition depends mainly on the original material. It contains almost all the essential nutrient elements in it. Leaf mould should be properly treated with fungicides and insecticides before use to kill the pathogens.

Compost

Compost is prepared by decomposing bulk organic wastes biologically under controlled conditions in piles or bins. Compost is a rich source of mineral nutrients with good water holding capacity. It can be used as a medium for propagation but should be mixed with soil. The various microorganisms like bacteria, fungi and nematodes and insects like beetles, cut worms, slugs can often be found in compost piles. Therefore, compost should be properly pasteurized before use.

Compost prepared from leaves have high salt contents, which can inhibit the growth of the propagules. It can be lowered down by leaching with water or mixing it with other propagating media.

Cocopeat

It is also called as coco dust. It is a byproduct of cutting and shifting of coconuts for fibre production. It is becoming very popular propagating and growing medium now-a-days because it has an excellent pore space (25-30 per cent) and fine structure required for proper growth and development of seedlings. Moreover, it is a rich source of nutrients and can easily be mixed with other growing media.

Rock wool

It is used as a rooting and growing medium in many advance countries of the world. It is derived from basalt rock, melted at a temperature 1600°C, then cooled and spun into fibres. Rook wool has good water holding capacity and retains good level of O_2.

Calcined clay

When minerals like clay, shades etc., are calcined (heated) at high temperature, they produce some stable aggregates. Such aggregates have no fertilizer value, but are porous, and resistant to break down and water absorption. Calcined clay and other aggregates are mainly used to mix with other growing media. Leca, Terragreon, Turfice, Haydate and Hortag are commercially used calcined aggragates as propagation and growing media in many advance countries like the USA, Japan, Germany etc.

Synthetic plastic aggregates

Synthetic plastic aggregates are also used in many countries, especially in Europe, as a substitute for sand or perlite growing media. The commonly used synthetic plastic aggregates are expanded polystyrene fractions and urea-formaldehyde foam. The polystyrene fractions are usually neutral, don't absorb water and don't decay. These usually improve drainage, aeration and bulk density when mixed with other media. Urea foams are usually sponge like structures having enough N content and high water holding capacity. The environmental problem of production and disposal of these materials limits their commercial use as propagating medium.

Shredded bark, wood shavings and saw dust

Shredded bark, saw dust and wood shavings of different plants like cedar, fir, pines, maple and redwood etc. can also be used in mixtures with various propagating medium. Because of low cost, lightweight and easy availability, these are mainly used in soil mixes for container grown plants. These mixes usually contain a lower amount of nutrients and hence additional amount of nutrients may be added to the mixes before their use as a growing medium. However, the shredded bark and wood shavings may contain toxic material like phenols, resin, terpentenes, and tannin etc., when fresh. Therefore, these should be composted for 10 to 15 weeks before use.

Soil mixes

Different soil mixes are used as a medium for propagation. Usually, soil mixtures are prepared by mixing sand, loam soil and leaf mould in different proportions. However, an ideal soil mixture should be porous and should have good water holding capacity. Soil mixtures are not only used as propagating medium for seed germination or rooting of cuttings but also for filling of containers. For example, in the USA, Cornell Peat-lite potting mix is used commercially. It refers to peat based media containing perlite or vermiculite in different proportions. However, the success of these mixes lies in the proportion and thorough mixing of the ingredients.

Raising plants in containers

Every day new techniques are developed and if appealing, these are taken up by the consumers. In the recent years, there had been great inclination of the plant lovers towards container grown plants. Accordingly, nurserymen are now widely adopting the technique of growing plants in containers. The container-grown plants have the following advantages:

1. Such plants are not frequently damaged during packing and long distance transport.
2. Fruit plants and ornamental foliage plants, which require greater care and management, are highly suitable for growing in containers.
3. Containers used for raising for plants can be reused repeatedly.
4. Most of the containers have higher durability.
5. The mortality rate of the container grown plants is very less.
6. Plants in container look attractive and help to fetch good price in the market.

Containers of different types, shapes and sizes are used by the nurserymen to raise plants. The chief characteristics of major types of containers used by the plant propagators are given below:

1. **Clay pots:** Clay pots are mainly used for raising some fruit and ornamental plants. These are available in different shapes and sizes but round type pots are avoided for raising plants because they cover more area in the nursery. The clay pots are porous in nature and thus permits frequent aeration and water movement. However, if continuously used, it is likely that pores of the walls are accumulated with salts, which may be harmful in later stages. These are mostly used in India.
2. **Rigid plastics:** Rigid plastic pots are also available in different shapes and sizes.

They are non-porous but occupy less space in the nursery. However, feasibility of their repeated use and light weight make their handling very easy during transportation.

3. **Fibre pots:** Fibre pots are manufactured from peat and wood fibre and are available in different shapes and sizes. These are biodegradable in nature and can be put in to soil along with plant. These are widely used in advanced countries but rarely in India.
4. **Polyethylene bags:** Polybags have replaced other containers (clay and rigid plastic pots) and these are now widely used by the plant propagators mainly because these are comparatively cheaper and can be used time and again. Moreover, handling of plants in polybags is very easy. These are light in weight and are easily available. Polybags of different sizes and thickness are available in the market, but in general, polybags of 30 x 20 cm size are used for raising fruit plants and 10 x 6 cm sized for raising seedlings of ornamentals and vegetables. However, before planting, these bags should be perforated with some needle for proper drainage.
5. **Coated paper containers :** These containers are also used widely in advance countries but rarely in India. These are light in weight and easy-to-handle during storage and transportation. These are manufactured from paper, cardboard fibre, impregmented with paraffin wax or bitumen or other non-absorbent material of non-poisonous nature. These are cheaper than the other containers.

Potting

Unlike fruit crops, which require large space to grow, ornamentals are mostly grown in pots for recreation purposes in houses. The plants are potted for the following reasons;

- Preparing plants for sale such as rooted cuttings of grapes
- Growing plants for decoration like crotons
- Growing plants for experimental studies like pot-culture studies
- For using plants as rootstocks in certain grafting methods as in inarching of mango.
- Pot mixture or potting compost
- It is essential for potting of plants.

Potting procedure

- Before lifting the plants from nursery, it is necessary to wet the beds so that root damage can be minimized. Care should be taken to lift the plant with a ball of earth with as much of the root system intact, as possible. It is advised that seedlings should not be pulled out in the hot sun besides not allowing roots or the soil around the roots to dry.

- Thereafter, fill up the pots by putting some crocks first, then a layer of sand (5-8 cm) and finally pot mixture (8-10 cm).
- Now, the plant can be placed with the ball of earth in the centre upon the layer of pot mixture (Place on one side of pots in the case of root stock plants used in inarching)
- Some pot mixture should also be put around the ball of earth then press as you fill up and level off, leaving one inch head space at top. Do not press over the ball of earth; otherwise, it will break and damage the roots.
- Set the stem of plant at the same height as it was in the seed bed.
- Immerse pot with plant in a tub of water gently and keep inside water till air bubbles cease to come out. Thereafter, remove and place the pot under shade of trees.

Repotting

Repotting is done for changing the soil medium for pot bound plants, which results when the potted plants are grown for more than one season or one year in pot and as a result the roots become a tangled mass very soon and exhaust all the nutrient in the limited soil, besides being circumscribed in the limited place. This stage is known as pot bound condition.

Repotting procedure

- It is always preferred to wet the potted plant 24 hours earlier to facilitate depotting (removal of plant from pot) followed by repotting (again placing in a pot).
- The technique to remove the plant with a ball of each intact is to keep the right hand palm over the soil, allowing the stem of the plant in between the first fingers and turn the pot upside down holding the pot at the bottom with the left hand and gently knocking the rim of the pot on the edge of table or any other hard surface or even on the bottom edge of another inverted pot. The ball of earth comes out of the pot. If for any reason, it fails to come out, break the pot knocking the sides with a stone or fork and free the soil from it.
- Examine the roots, cut neatly with a secateur, the decayed, dead and dried or twisted roots. Reduce the size of the ball of earth around the roots.
- Place the plant in the new pot at the same height at which it was in the old pot. Fill up pot with fresh pot mixture and immerse in water.

***

14 Structures Used in Plant Propagation

For proper growth and development, plants require optimum supply of light, water, temperature, gases and mineral nutrients. Thus, for successful propagation and growing of plants, these basic necessities should be available or created artificially to get optimum results. In order to enhance the efficiency of propagation of plants, commercial producers manipulate the environment of propagation area or locality or place by managing (1) Micro-climatic conditions like relative humidity, temperature, light and gases etc. (2) Edaphic factors i.e., propagation medium or soil, mineral nutrients and water etc., and (3) Biotic factors i.e., interaction of propagules with insect-pest and pathogens. For controlled environment or micro-climate, various structures and equipments have been developed, which are used by the propagators and nurserymen and if handled properly, they maximize the plant growth and development by providing the best environment for the growing plants. The different propagation structures used for plant propagation are greenhouse, hot bed, cold frame and lath house. Greenhouse and hot bed are primarily used for providing the controlled environmental conditions for seed germination or cuttings, whereas, the lathhouses and cold frames are used for hardening of young plants before transplanting. In addition, many other structures are also used, which may be of some significance for commercial propagators of advance countries. A brief description of different propagation structures and equipments has been given below:

Greenhouses

A greenhouse greatly extends the variety and scope of propagation. This is important, particularly where most of the propagation is done in late winter and early spring. The framework supports should be as slender as the necessity for a certain degree of strength and stability in the structure for light obstruction. Steel and aluminum fulfill these conditions best, but are usually more expensive than timber. Experiments have shown that a greenhouse that runs from east-to-west is best for better light penetration in winter and early spring, and consequently preferable for raising seedlings at this time of the year. Moreover, it is important that the greenhouse should be well away from any kind of shade such as a tree or buildings, including other greenhouse. Some shelter, however, from north to northeast winds is desirable. These glasshouses may be plastic polyethylene covered or made from fiberglass. Modern green houses are well equipped with elaborate structures and have precise control on temperature, light intensity and humidity. The size and type of greenhouse, primarily depends upon the need of the plant propagator. Glass covered greenhouses are expensive but they have long life. However, for short-term benefits, plastic covered greenhouses can also be made. Two types of plastics are available for their construction, polyethylene and fiberglass. Both are lightweight and inexpensive as compared to glass. Polyvinyl chloride (PVC) has also been used in construction of greenhouses but it tends to darken pre-maturely in sunlight.

In India, construction of temporarily low-cost polyhouses is in fashion for raising nursery of fruit plant in off-season. Such low-cost greenhouses are constructed either on wood or metal framework and are covered with polyethylene sheet of 0.10 to 0.15mm thickness, which is resistant to ultra-violet rays. These houses are equipped with thermostat, cooler or an air conditioner or humidifier etc., for rigid control on temperature and humidity. Greenhouses made from fiberglass sheet are more durable than the polyethylene houses, but are quite expensive.

Hot bed

The hot beds are small low structures, used for propagation of nursery plants under controlled conditions. Usually, hot bed consists of a frame, the cover and a heating system. The frame is made up wood, steel or aluminum, and covered with a plastic or glass, while heat is provided artificially below the propagating medium by hot air or water. A thermostat automatically controls the temperature. Hot beds are widely used for raising plants in western countries but in India these are rarely used as the cost involved in construction is very high. As in greenhouse, close attention must be paid in hot beds to shading and ventilation and on temperature and humidity control. For better results, hot beds should be placed in sunny location but in protected and well-drained location. Hot beds can be used throughout the year, except in areas with severe winter, where their use may be restricted to spring, summer and fall.

Cold frame

Cold frames in the open have many advantages in propagation. Softwood cuttings from herbaceous plants and shrubs can be raised in them during warmer months. In autumn, semi-mature or hardwood cuttings may be planted in cold frames. Frames are also useful for starting early vegetable crops and other plants from seeds. Cold frames can be used for the conditioning or hardening of the rooted cuttings, young seedlings or container plantings and starting new plantings in late spring, summer or fall, when external heat supply is required. Sowing can be done under the continuous type of cloche to give seedling an early start. These are similar to hot beds except that there is no provision for supply of bottom heat. In cold frame, heat is provided through sun only and hence these are kept in sunny places only. For their construction, the standard glass (0.9 x 1.8 m) is often used as a covering for the frame but less expensive polyethylene or fiberglass covering can also be used. The covered frames should fit tightly to retain heat and humidity. Cold frames should be placed in sunny locations, protected from winds and should have covering sloping down from north-to-south. To get desired results, a careful attention should be paid on ventilation, shading and winter protection in cold frames.

Lathhouse

Lathhouse provides outdoor shade and protects container-grown plants from high summer temperature and high light intensity. They may reduce moisture stress and decrease the water requirement of the plants and thus these are useful for protecting young seedlings or rooted cuttings or container plants from high summer temperature and light intensities. Lathhouse serves as an intermediate between cold frame, green house or a hot bed. The construction of a lathhouse depends on the need of the propagator. These can be made from pre-fabricated aluminum or wood but aluminum made structures are costly. Mostly, wood or pipe supports are used, which are set in concrete with necessary supports. For shading, woven plastic material is widely used to allow proper light inside the house. In summers, however, overhead sprinklers are more useful than shade for cooling down the temperature. Water requirement of plants grown under lathhouse is usually low because of low rates of evaporation and transpiration.

Propagation frames

Sometimes, in greenhouses, the humidity is not enough to allow satisfactory rooting in the leaf cuttings. In such cases, enclosed frames covered with glass or plastic material may be used for rooting of cuttings. These frames are useful only in grafted plants as these retain high humidity during the process of healing. Large inverted glass can also be kept over a container having cuttings. Though, high humidity is required in such frames but proper ventilation and shading is necessary after the rooting process has started in the cuttings. Warm and humid conditions inside these structures provide excellent environment for growth of pathogenic organisms, which may infect the propagating material. It is, therefore, necessary to maintain cleanliness and proper sanitation in these structures.

Net house

Net house is widely used as propagation structure in tropical areas, where artificial heating is not required and artificial cooling is expensive. In these areas, a net house may be constructed with roofs covered with glass or plastic film and its sides are covered with wire net. It provides necessary ventilation and maintains an ideal temperature for germination of seeds and subsequent growth of the seedlings. The roof of net house may be covered with gunny cloth or even with live plant creeper to cut off the solar radiant energy and to keep the house cool. Net house can be constructed as per the need of the propagator and therefore its size varies with the requirement of the nurserymen.

Bottom heat box

It is a simple device for promoting rooting of cutting in difficult-to-root fruit plants like mango and guava. It consists of two chambers made from galvanized iron sheets. The outer chamber has a height of 70 cm with 46 cm width and the inner chamber has a height of 68 cm and width of 44 cm. The space between the two chambers is filled with glass wool for heat insulation. The another chamber is fitted inside the inner chamber with the height of 35 cm, fitted with two electric bulbs at the bottom for providing heat to the cuttings. The innermost chamber is filled with soil mixture or any other propagation media and the cuttings are inserted in it. Two electric bulbs of 100-Watt capacity are fitted at the bottom of the chamber to provide heat and light to the cuttings. Similarly, the temperature in the box is maintained and controlled automatically by thermostat fitted at the bottom of the chamber. The most ideal temperature to be maintained in the box is 30$\pm$ 2°C because at this temperature, cuttings of mango and guava root easily and profusely. The initiation of rooting in cuttings varies from species to species but in general, it takes 1-2 months for proper development of the roots.

Forsythe Pot

A forsythe pot is often recommended for home propagation. It is made with two clay pots, one about 2 inches in diameter, the other 6 to 8 inches in diameter. The rooting medium is placed in the larger pot. Plug the drainage hole of the small pot and set it in the center of the larger one and level with the edge. The small pot is filled with water that seeps into the rooting medium where the cuttings have been inserted.

Terrariums

Terrariums or small aquariums are also good or one can make his own glass or plastic covered propagation box. A plastic bag or glass jar placed over the cutting will increase the humidity or you can mist the cutting with water several times a day. Plants with thicker or waxy leaves suffer less from low humidity than do thin-leaved plants. Plants in sealed containers may require a fungicide spray to deter harmful pathogens.

Mist propagation unit

The rooting of soft wood leafy cutting under spray or mist is a technique now widely used by nurserymen and other plant propagators throughout the world. The aim of misting is to maintain continuously a film of water on the leaves, thus reducing transpiration and keeping the cuttings turgid until rooting takes place. In this way, leafy cuttings can be fully exposed to light and air because humidity remains high and prevents damage even from bright sunshine.

Mist propagation units are used for propagation of "difficult-to-root" cuttings in most advanced countries. Mist beds are constructed within a greenhouse. A fine mist is sprayed intermittently over the cuttings at regular intervals during day and night. The mist unit is controlled by a time clock, operating a magnetic solenoid valve and is set in a way to turn on the mist for 3-5 seconds to wet the leaves and turn off for some time and when the leaves are dry, the mist is again turned on.

Plastic tunnels

A simple but effective method of protecting cuttings has been developed by Greenhouse Crops Research Institute at Littlehampton, England, which is being used by many nurserymen throughout the world. The three feet wide wire hoops, which support the polyethylene are made from 0.2-inch diameter wire and are erected at 30-inch intervals. A white translucent polyethylene sheet 6 feet wide is then stretched over the hoops and is held in position by fastening two lines of polyethylene bailer twin to alternate hoops on either side of the tunnel (**Plate 8**). The seeds or cuttings or other propagating material is sown/planted inside the tunnels during winter. Nowadays, perforated tunnels are used because in non-perforated tunnels, temperature and relative humidity rises sharply, which may affect rooting of cuttings or further growth of the propagating material.

Propagating units

There is now wide range of self contained propagators, heated and unheated, in the market. The simplest type consists of nothing more than a standard plastic seed tray with a transparent cover known as a propagator top, which is made to fit over the tray. The propagator tops are made from clear polystyrene and are fitted with one or two adjustable ventilators.

Larger and more sophisticated propagators are also available. One of these is rather like a glass case with a strong metal framework. It is nearly 3 feet long, 2 feet wide and 2 feet high. This propagator is fitted with sliding glass panels both front and rear making for easy access. Similar models are available in small sizes and each has a damp-proof base about 6 inch deep.

Another excellent type of propagator is rather like a miniature greenhouse with sliding panes both on the sides and the roof to provide access and ventilation. A model of this type has an aluminum framework and a fiber glass base. Its dimensions are 4 feet 2 inch long by 2 feet 2 inch wide and 2 feet 9 inch high at the ridge.

Most propagators are provided with heating equipment either with bottom heat or fitted with an air heating system or both air and bottom heat is provided but electrical heating is considered the best. However, a gardener could easily make a simple type of propagator for himself. To do this, make a sand bed at the base of a deep box with bottom heat and provide a layer of about 3-inch of rooting compost. Over the bed, erect a light framework to carry a plastic sheet. A deeper box could have simply a sheet of glass laid over as a cover.

Growing rooms

A growing room is an insulated building from which natural light is usually excluded. In it, illumination is provided by artificial means. Growing room are now widely used commercially for the production of seedlings such as bedding plants, tomatoes and cucumbers in most advanced countries. The seedlings are usually grown in trays or pots kept on benches. To save space, the benches are usually installed in tiers being vertically about 2 feet wide and 6 inch apart. For convenience, they are usually 8 feet long by 3 feet 6 inch wide. Each bench is illuminated with 8 feet long 125-watt fluorescent tubes mounted 1 feet 6 inch above the bench. Seven tubes over each bench provide a light intensity of 500 lumens per square feet, which is adequate for bedding plants. However, if plants grown in such houses have higher requirement for light, more fluorescent tubes can be installed over the benches containing the plants. The heat of the tubes usually maintains a temperature of at least 70°F and excessive temperatures are prevented by use of fans installed in building.

The automatic greenhouse

Today, the modern greenhouse can be almost completely automated thus assisting propagation. For instance, by the use of thermostats, air and bed temperatures can be maintained at as per the requirement. Similarly, automatic ventilation allows the ventilators to open and close in relation to temperature. Even, automatic systems of irrigation are installed in the modern greenhouses and water is supplied to the plants through drip or trickle system to each pot or plant by individual nozzle or tube. This system is controlled by automatic "on" and "off" valves or by means of a time switch.

Most advancement in raising of plants under highly controlled environmental conditions has been the use of phytotron facility. National Phytotron Facility is available at IARI, New Delhi

* * *

Tools, Equipments and Accessories

In order to have excellent output from nursery one should have to be careful in performing several operations in time with expertise. In day-to-day routine, various cultural operations are carried out in the nursery. For this purpose, many tools and accessories are required from time to time. Some tools are simple and are used for simple operations. However, special equipments are required for specific operations. Some of the tools and accessories required for different operations in the nursery are described below:

Knives: In general, in the nursery, two types of knives *viz.* budding and grafting knives are required for propagation work. A grafting knife in general has a straight 7.5 cm long blade and a strong long handle. The budding knife, on the other hand, may have a straight or a bit curved blade of shorter length. It has, however, a spatula at the end of a handle, which is used for lifting the bark during the budding operation. Sometimes a knife having two parallel blades (double bladed knife) is also necessary. It is specially required for lifting or removing the patch of a bud from the budwood. Similarly, a patch budder tool, having two pairs of blade is specially required for patch budding. However, very often a budding knife can be used for grafting purpose also. These knives may have either a folding or fixed blade. The blades of the knives should be made from high carbon steel and should always be very sharp.

Kudali: It is an important tool for digging of pits or digging of soil. Kudali has a metal blade attached to a wooden handle. It is operated manually and considered very important nursery tool.

Garden fork: Garden fork is mainly used for breaking of soil clods during preparation of nursery beds or fields.

Weeding fork: It loosens the soil and is equally useful in weeding. It consists of a long handle with a blade having teeth. It is drawn manually with the help of a handle to collect the weeds and cops of plants etc.

Khurpi: Khurpi is widely used for various cultural operations in the nursery but mainly for removal of weeds and loosening of soil for proper aeration. It has a sharp edge, triangular blade fitted with a wooden handle.

Crow bar: It is an iron rod with one end pointed and other as a wedge shaped. Crow bar is mainly used for digging of pits.

Water cans: Watering cans are used for irrigation for irrigation of annual flowers, vegetables and newly transplanted seedlings of various flowering annuals and vegetables or young plants in pots. These are made up of galvanized iron sheet and sometimes of plastic. These cans are fitted with a rose over the nozzle, which is very useful for equal distribution of water over the germinating seedlings.

Garden rakes: Garden rake is used for collecting stones and bricks bits from the bed, scarifying the grass surface and gathering the fallen leaves. A toothed rake is always better for such operations.

Secateurs: Secateur is considered as the most important tool for a propagator or a nurseryman. Secateur is used for excising scions, lopping off the rootstock, removing the undesirable sprouts/shoots from the stock, preparation of the scion sticks and for pruning operation. The blade of the secateur should be of a good quality because poor quality blades may not give smooth cuts to the stock and scion.

Grafting machines: Many machines have been developed and commercially used for the preparation of scion sticks and bud wood for budding and grafting operations in many developed countries of the world. These machines are, however, not very common in India.

Pruning saws: Several types of saws are required for different operations involved in propagation of horticultural plants. The commonly used are crescent saws, tapered saws and straight saws. All of them have long and widely set teeth to facilitate pruning or cutting of green wood. In general, a small saw is required for cutting the hard branches of the stock plant and sometimes the same saw can be used for pruning also. Its blade should be narrow so that it can pass through the narrow or closely spaced branches.

Pruning shears: The different types of shears like hand shears, lopping shears, tree trimmers etc., are needed in a garden. Pruning shears should not be very expensive but these should be made up of good steel. Similarly, these should make a smooth and clean cut with least injury to the plant.

Ladder: In propagation work, ladder is required for operations like cutting of bud wood, training of vigorous plants and top working of old and declining plants. In general, step ladder or straight ladders or hook ladders are used for such operations in the propagation field or nursery.

Tying and wrapping material: In most of the grafting and budding methods, it is essential to hold scion and stock firmly together to have successful graft/bud union, for which, a suitable tying or wrapping material is required. Accordingly, nurserymen and other propagators use different materials to achieve desired results from budding and grafting.

a. **Waxed string and cloth:** It is the most commonly used tying material by the nurserymen. It can easily be prepared by dipping narrow strips (25-30 cm wide) of long cloth in the melted bee wax. These waxed cloth strips can be used for tying the stock and scion after budding or grafting. The added advantage of using these stripes is that these eliminate the need of further waxing. Such type of wrapping material restricts the entry of water to the bud union.

b. **Waxed string:** Next to waxed cloth, waxed strings are the most commonly used tying material. It can be prepared by soaking a ball of cotton twine in a mixture of melted resin (I Kg), bee wax (500 g), and linseed oil (250 g) and paraffin (120 g) for 10-15 minutes. This cotton ball should be rotated repeatedly while kept in the above mixture. For large scale grafting operation, waxed string is most convenient because it adheres to plant parts without tying. However, it should be strong enough to hold the grafted plant parts together and week enough to be broken by hand.

c. **Raffia fibre:** It is a strong fibre and is not easily damaged by heat or rain. For proper working with raffia fibre, it must be moistened by wrapping in damp cloth overnight.

d. **Rubber strips:** Rubber strips provide better grip to the stock and scion than raffia fibre or waxed cloth. These are highly elastic and can be stretched to a great extent. The greatest advantage of these strips is that these don't girdle the plant even if these are not removed after the union has taken place.

e. **Polythene strips:** Plastic strips of different thickness are used as a tying material in budding and grafting by the nurserymen. In India, plastic films are more common amongst nurserymen. These are slightly elastic and don't girdle the plant and commercially used by nurserymen.

In addition to the above mentioned tying material, adhesive tapes, similar to surgical adhesive tapes but lighter in weight, are also used by commercial nurserymen. Similarly, paraffin tape is also used by some propagators.

Baskets: A gardener should have plenty of baskets. These baskets are mainly used for carrying the seedlings for packing and transportation to the distant places. Wooden baskets are very common in India

Labels: A commercial propagator should have different types of labels. These labels are used for proper labeling of plants before sale. The labels may be made up of paper, cardboard, wooden, celluloid, aluminium and plastic etc. but usually the plastic or aluminium labels are preferred. However, for temporary use, the paper or cardboard labels can be used. In advance countries, plastic labels are widely used.

Pots: Pots of different shapes and sizes are used in the nursery. The clay as well as plastic pots are available in different shapes and sizes. Usually, 4" (10cm) pots are suitable for small annual seedlings, six inches (15 cm) for palms and ferns, 8 inches (20 cm) for single specimens. In India, earthen pots are mostly used but now there is trend towards plastic pots because plastic pots are durable and are easy to handle.

Seed pans: Seed pans are usually smaller pots, which are used for raising seedlings of annual vegetables and ornamental plants. The seed pans are usually flat with size varying from 30 to 50 cm in diameter.

Seedling trays: These are primarily used for hardening the micropropagated plants and for germination of seeds of different horticultural crops. The size of these trays varies with the number of cavities, ranging from 25 to 250. The propagators may chose the seedling trays of different capacity or as per their need and the crop to be raised.

Planting supports: Bamboo of different sizes is required for staking the plants during their initial stages of growth. These should be available with the person-raising nursery. In place of bamboo, iron rods can also be used.

Grafting wax: Grafting wax is used by nurserymen at least with two objectives i) To seal the graft union for preventing the moisture loss and desiccation and death of the cells of the cut surfaces, which are essential for healing of the graft union, and ii) To prevent the decay of wood by way of checking the entry of the pathogens.

An ideal grafting wax should have the following characteristics

1. It should adhere well to the stock and scion.
2. It should not get washed off easily by the rains.
3. It should not be so brittle to crack during cold weather.
4. It should not be very soft to melt and run off during hot weather.
5. It should be pliable enough to allow the expansion of the stock and scion without cracking.

The commonly used waxes are either hot or cold.

a. Hot wax: Hot wax can be prepared easily at home without involving much cost, labour or instrument. The ingredients used in preparing hot wax are resins, bee-wax, raw linseed oil, powdered charcoal and some glue. The proportion of these ingredients may vary but the most commonly used proportions of each ingredient is given below:

Resin	=	2.25 Kg
Bee-wax	=	0.40 Kg
Raw linseed oil	=	240 ml
Powdered charcoal	=	40 g

First of all, the glue is boiled in sufficient quantity of water in container and resin and bee-wax is boiled in an another container. Both the ingredients are allowed to cool. Now add linseed oil and powdered charcoal to this mixture. The glue is then added slowly and the mixture is stirred continuously. Pour out this mixture into shallow greased pans and wooden box lined with a greased paper. It is then allowed to harden. For use, cut a small piece, reheat it in a container and apply with the help of a paintbrush.

b. Cold wax: Cold wax chiefly contains an emulsion of asphalt and water. In some instances, it has given satisfactory results. The cold wax chiefly contains water (50%) and after application, the water evaporates, leaving the coat of asphalt over the graft. Thus, if rains occur with 24 hours of its application, the rains may wash it off and hence re-waxing of the graft union is necessary. Further, these waxes are easily broken down by freezing, hence its containers should be kept in warm places during winter.

Waxed tape: It is prepared by tearing a domestic cloth in to bands. These bands are rolled and dipped in the hot solution of resin, bee wax, tallow, paraffin and linseed oil. The waxed strip of 1/3 to 1/2" in width is cut cross wise from the roll as per need. For good life of the tape, a good grade domestic cloth should be used.

Waxed patches: Waxed pathces are prepared by tearing the domestic cloth with a hot mixture of resin, bee wax, tallow, paraffin and linseed oil. Afterwards, the square or rectangular patches of 1.5 to 2" size are cut with a hole in the centre. These patches are placed on the graft or bud union in a way so that bud or scion sprouts through the whole. Such patches are used widely in patch and ring budding.

Cotton twine: It is popular tying material for buds and grafts on small stocks. 10-12 ply twine is usually used for tying grafts and buds on vigorous rootstocks and a bit smaller for dwarfing rootstcoks. A good grade twine should always be used to avoid injuries to the tissues at or near the graft or bud union.

Protective materials: The wounds caused by pruning or cutting of scion, bud-wood may act as a source of secondary infection by the pathogens and insects. It is, therefore, advisable to cover the wounds with some protective material. The protective material should have the following characteristics:

1. It should give full protection against every type of infection.
2. It should adhere properly to the wounded surface.

3. It should neither crack nor run the surface after application.
4. It should not be injurious to the injured tissues.

Bordeaux paste is the best material for this purpose. However, blitox can also be used as an effective protective material.

Pesticides and fertilizers: A propagator should have all chemicals to control insect-pests and disease infestation in the nursery. Similarly, for adequate nutrition of the plants, manures and fertilizers should be available at all times. Rats cause great damage to the nursery plants and therefore repellents or poison baits should also be kept ready every time.

***

16 Formulations of Growth Regulators (Powders, Solutions and Pastes)

Before discussing about formulations of growth regulators, it is very important to have clear understanding about the terminologies used in this context. Growth regulator is a broad term and all growth regulating compounds either endogenously or exogenously produced comes under this category. In relation to plants, terms like hormones or phytohormones are synonymous to each other. Both defines hormones produced by plant system. Plant hormones are organic compounds, other than nutrients, produced by plants, which in low concentrations regulate plant physiological processes. They usually move with in the plant from a site of production to site of action. Plant growth regulators are either synthetic compounds or plant hormones that modify plant physiolosical processes. They regulate growth by mimicking hormones.

In fact, horticulture industry is indirectly dependent on plant growth regulators as these are exclusively used for different purposes in all stages of the plant growth for desired manipulations like treatment of seeds for enhancing their germination, rooting, flowering, fruiting, yield, ripening and as post harvest treatments for better shelf-life. Further, for propagation of horticultural plants, different growth regulators are also used for the treatment of cuttage and layers for inducing roots in them. These are applied as powders, solutions or as a paste. Though, some ready-made commercial products are also available in the market. However, one should be able to make the solutions or pastes at home. A brief procedure for the preparation of hormonal powders, solutions and pastes is given in the following pages. In general, it should be kept in mind that one milligram pure hormone is required in one kilogram of inert material (talc etc.) or distilled water or lanolin for making one ppm strength of that particular hormone or growth regulator. Similarly, it is necessary to remember that most of the hormones are not easily dissolved in water. Therefore, it is necessary to dissolve them first in a suitable solvent. As a principle if growth regulators are of basic reaction then it should be dissolved in the solvent having acidic reaction. For this, the required quantity of the hormone is weighed and dissolved in a few drops of alcohol or other solvent, and then shaken until it dissolves completely. Then, the final volume should be made with distilled water. Another thing, which is of practical importance is that these growth regulators are required in a very minute quantity, therefore, precise weighing is of utmost importance and any mistake can be proved fatal for plants.

Preparation of hormonal powders, solutions and pastes

A. **Hormonal powders:** For preparing hormonal powders, the required quantity of the hormone is weighed precisely with the help of sensitive balance. It is dissolved in 1/2 litre ethanol, methanol or acetone in a beaker. This material is poured in to one kilogram of talc taken in mortar and mixed thoroughly with a glass rod. After mixing, the mixture is kept open in air for few hours. The alcohol will evaporate soon, after which, the dried talk is ground to a fine powder. This fine powder should be kept in airtight containers to avoid moistening and can be used as and when required.

B. **Hormonal solutions:** For the preparation of hormonal solution, the required quantity of the hormone is weighed accurately in a physical or electronic balance. It is then dissolved in a few drops of acetone or alcohol. When the contents are fully dissolved, the final volume is made with distilled water. The pH of the solution should be nearly neutral or so. If the pH of the solution would be high or less, it may not serve the purpose, rather it would affect the process adversely. The pH of the solution can be adjusted with the help of 0.1N HCl or 0.1 N KOH. Hormonal solutions should preferably be prepared fresh. These can, however, be stored at low temperature in a refrigerator for some time but stored preparations may loose their properties and become inactive.

C. **Hormonal pastes:** For preparing hormonal pastes, the required quantity of the hormone is weighed accurately and dissolved completely in a few drops of alcohol. The required quantity of the lanolin (wool fat, a product similar to grease and have greenish-yellow in colour) is also weighed and heated slightly in a beaker under gentle flame. When, the lanolin is slightly liquidified, the dissolved hormone is poured in it. The contents are dissolved thoroughly with constant stirring with a glass rod. The contents are then allowed to cool down. The paste is thus ready for use. Until use, the paste may be kept for few months in a cool dry place but one should prefer to use fresh paste.

Precautions

The use of root promoting hormones for induction of rooting in the cuttings and layers has now become the backbone of the nurserymen. These are readily available in the market in various forms but it is always better to prepare them at nursery. The hormones not only induce rooting but also help in growth and development of the roots. IBA and NAA are most effective in cuttage and layers and these are used in tissue culture systems also. To have better and desired results from growth regulators, the following precautions should be followed for their preparation and further use:

- First of all, check for expiry date of the hormonal powder.
- Hormones/ growth regulators should be weighed precisely and accurately, preferably on an electronic balance.
- As most of them don't dissolve in water, a proper solvent (acetone/alcohol etc.) should be used.
- Only a minute quantity is used at once. So don't buy more than what is required or nearly so.
- Hormones are liable to deteriorate soon if kept under high temperature. Therefore, keep them in a cool and dry place, preferably in a refrigerator.
- Few hormones are photosensitive and photodegradation occurs if exposed to light as in IAA (Indole Ascetic Acid). Therefore, these should be stored in dark or amber coloured bottles.
- Nurserymen should have fair knowledge about concentrations of specific growth regulators for specific fruit species and specific purpose.
- Solutions should always be made fresh. If required to store for some time, use of refrigerator is desirable.
- Use hormonal solutions for treatment of cuttings and lanolin paste for layers.

Methods of application of growth regulators

As stated earlier, plant growth regulators are widely used as an effective aid for rooting of layers and cuttings. Among different plant growth regulators, auxins are most effective and widely used as a rooting aid. Among synthetic auxins, IBA and NAA are found to be more effective than the others. However, 2,4-D, 2,4,5-T, 2,4,5-TP etc., are also effective in inducing roots in some plant species if used at low concentrations.

In general, the effectiveness of plant growth regulators is not only dependent on the concentration, type of the plant species but also on the method of application. Different methods are used for the treatment of cuttings and layers with plant growth regulators. These are described briefly hereunder:

a. **Prolonged soaking method:** In this method, the basal ends of cuttings are dipped in the dilute solution of the hormone for 24 hours in a cool dry place. After the treatment, the cuttings are planted in the nursery or in other suitable growing medium. The concentration of the hormone or growth regulator usually varies from 20 ppm to 200 ppm, depending on the plant species and type of the cuttings. The concentration is usually low for easy-to-root species and *vice versa*. The nurserymen rarely use this method of treatment because it is a bit cumbersome and laborious method. However, it is very useful for difficult-to-root species, where some materials like vitamins, sugars and nitrogenous compounds are also used along with the growth regulators for facilitating rooting in such species.

b. **Quick dip method:** This method is an improved version of the previous method. It is very effective method of treating cuttings with growth regulating chemicals and is usually followed by most of the plant propagators for raising plants through cuttings. In this method, the basal ends of cuttings are dipped in the concentrated solutions of a hormone for a short time, usually for 5 seconds to 2 minutes. The treated cuttings are then planted in the nursery or field for rooting. The concentration of hormone for quick dip method may be 500 ppm to 10,000 ppm, depending on the species and type of the cuttings but usually a concentration of 4,000 to 5,000 ppm is used. In general, the higher concentration should be avoided as it may cause injury to the cuttings.

c. **Powder dip method:** In this method, the basal ends of freshly prepared cuttings are dipped in the hormonal powder carrier (talc) for some time. After treating the cuttings, extra powder adhering to the cuttings should be removed by shaking and cuttings are immediately inserted in the rooting medium. Seradix, a popular formulation is used by the nurserymen by this method. For effective rooting, the cut ends of the cuttings should be moistened before the treatment and at the same time, the excess of powder should be shaken off, otherwise it may cause adverse effect of the rooting process.

d. **Spray method:** Spraying of growth regulators is sometimes done to the mother plants before taking cuttings from them. Spraying of stock plants with 2,4,5-TP in concentrations ranging from 25 to 100 ppm is done about 30 to 40 days before taking cuttings from them. Cuttings taken from such plants root better as compared to un-treated plants.

e. **Lanolin paste method:** As described earlier, some growth regulators (especially IBA) are applied to the girdled portion of a layer or stool in lanolin paste for inducing rooting in them.

* * *

Bibliography

Abbott, A.J. and Atkin, P.K. (1987). Improving vegetatively propagated crops. Academic press, New York.

Adams, C.R., Brandford, K.M. and Early, M.P. (1993). Principles of Horticulture. Butterworth Heinemann Ltd. Jordan Hills, Oxford.

Ahuja, M.R. and Libby, W.J. (1993). Clonal forestry. Springer-Verlag, Heidelberg, Berlin.

Altman, A. (1998). Agricultural Biotechnology. Marcel, Dakker, Inc, New York.

Bal, J.S. (1997). Fruit growing. Kalyani Publishers, Ludhiana, India.

Bendre, A.M. and Pande, P.C. (2002). Introductory Botany. Rastogi Publications, Shivaji Road, Meerut-250 002.

Bewley, J.D. and Black, M. (1994). Seeds: Physiology of development and germination. Plenum press, New York.

Bhojwani, S. S. and Bhatnagar, S.P. (2000). The embryology of angiosperm. 4th revised edition, Vikash Publishing House Pvt. Ltd., New Delhi-110 014.

Bose, T.K., Mitra, S.K. and Sanyal, D. (2001). Fruits: Tropical and Subtropical. Naya Udyog Publishers, Calcutta.

Bose, T.K., Sanyal, D. and Sadhu, M.K. (1998). Propagation of Horticultural Crops. Naya Prokash Publishers, Calcutta.

Bose, T.K., Som, M.G. and Kabir, J.(1993). Vegetable crops. Naya Prokash Publishers, Calcutta.

Brison, F.R. and Adriance, G.W. (2000). Propagation of horticultural plants. Biotech Books, New Delhi.

Bunt, A.C. (1976). Modern potting composts: a manual on the preparation and use of growing media for pot plants. University Park, Pennsylvania State University.

Carlson, R.F. and Mitchell, A.E. (1971). Budding and grafting fruit trees. *Michigan Agri. Expt. Bull.*, 508.

Chadha, K.L. (2001). Handbook of Horticulture. Indian Council of Agricultural Research, New Delhi.

Chattopadhyay, T.K. (1999). A text book on pomology. Vol. I : Fundamentals of fruit producrtion. Kalyani Publishers, Ludhiana, India.

Conger, B.V. (1981). Cloning agricultural plants via *in vitro* techniques. CRC Press, Florida, USA.

Corner, E.J.H. (1976). The seeds of dicotyledons. Canbridge University press, Cambridge.

Davidson, H. and Mecklenburg, R. (1981). Nursery management: Administration and culture. Englewood

Cliffs, Prentice Hall, New York.

Denisen, E.L. (1979). Principles of Horticulture. 2nd edition, Macmillan Publishing Company, New York.

Dijk, P.V. and Damme, JV. 2000. Apomixis technology and the paradox of sex. Trends in plant Science, 5(2): 81-85.

Dutta, A. C. and Dutta, T. C. (2000). Boatany, 6th revised edition, Oxford University Press, Culcutta.

Edmond, J.B., Senn, T.L., Andrews, F.S. and Halfacre, R.G. (1977). Fundamentals of Horticulture. Tata Mac-Graw-Hill Publishing Company Ltd., New Delhi.

Fahn, A. (1982). Plant anatomy. Pergamon press, New York.

Finch-Savage, W.E. and Gerhard L.M. 2006. Seed dormancy and the control of germination. New Phytologist, 171: 501–523.

Garner, R.J. (1979). The grafter's handbook. 4th ed., Oxford University Press, New York.

Garner, V.R. (1986). Principles of horticultural production. Michigam State University Press, East Lansing.

Geneve, R.L. (1991). Pattern of adventitious root formation in English ivy. *J. Plant Growth Regul.,* **10**: 295-220.

Gill, S.S., Bal, J.S. and Sandhu, A.S. (1985). Raising fruit nursery. Kalyani Publishers, Ludhiana, India.

Hartmann, H. T. and Beutel, A. (1979). Propagation of temperate zone fruit plants. Leaflet, California, Agri. Expt. Sta., Calfornia.

Hartmann, H. T. and Kester, D.E. (1986). Plant propagation. Prentice Hall of India Pvt. Limited, New Delhi.

Hartmann, H. T., Flocker, W.J. and Kofranek, A.M. (1981). Plant Science: Growth, development, and utilization of cultivated plants. Englewood Cliffs, N.J. : Prentice Hall, New York.

Hartmann, H.T., Kester, D.E., Davies, F.T. and Geneve, R.L. (1997). Plant propagation: principles and practices. Prentice Hall of India Pvt. Ltd. , New Delhi.

Hayes, W.B. (1960). Fruit growing in India. Kitabistan, Allahabad.

Hilhorst, H.W.M. and Koornneef, M. 2007. Dormancy in Plants. Encyclopedia of Life Sciences. 1-4.

International Seed Testing Association (1978). International rules for seed testing. *Seed Sci. and Technol.,* **4**: 114-177.

Justice, O.L. (1972). Essentials of seed testing. *In:* Seed Bilogy. Vol. 3., T.T. Kizlowski, ed. Academic press, New York.

Khan, A.A. (1981). Hormonal regulation of primary and secondary dormancy. *Israel J. Bot.,* **29**: 207-224.

Kucera B, Cohn MA, Leubner-Metzger G. 2005. Plant hormone interactions during seed dormancy release and germination. Seed Science Research 15: 281–307.

Leopard, A.C. and Kriedmann, P.E. (1975). Plant growth and development. 2nd edition, McGraw-Hill Company, New York.

Lovel, P.H. and White, J. (1986). Anatomical changes during adventitious root formation in fruit plants. *In:* New root formation in plants and cuttings. Martinus Nijhoff Publishers, Dordrew, The Netherlands.

Malik, M.N. (2000). Horticulture. Biotech books, New Delhi

Mascarenhas, A.F. (1997). Handbook of plant tissue culture, ICAR, New Delhi.

Mayer, A.M. (1981). Control mechanism in seed germination, *Israel J. Bot.,* **29** : 1-4.

Mcmillan-Browse, P. (1979). Plant propagation. Simon and Schuster, New York.

McMillan-Browse, P.D.A. (1978). Stratification - a detail of technique. *Proc. Intl. Plant Pro. Soc.,* **28**: 191-192.

McMillan-Browse, P.D.A. (1979). Plant propagation. Simon and Schuster, New York.

Mengel, K. and Kirkby, E.A. (1987). Principles of plant nutrition. 4th edition. International Potash Institute, Worblaufen-Bern, Switzerland.

Moore, R. and Walker, D.B. (1981). Graft compatibility-incompatibility in plants. *BioScience,* **31(5)**: 389-391.

Mukherjee, S.K. and Majumder, P.K. (1986). Propagation of tropical and Sub-tropical fruit crops. ICAR, New Delhi.

Nelson-Jones, W. (1969). Plant chimeras. Methuen and Company, London.

Nickell, L.G. (1982). Plant growth regulators: agricultural uses. Springer Verlag, New York.

Nygren, A. (1954). Apomixis in angiosperms II. *Bot. Rev.,* **20**: 577-649.

Pandey, S. N. and Chadha, A. (2000). A text book of botany: Plant anatomy and economic botany, Vol. III, Vikash Publishing House, New Delhi-110 002.

Pike, A.V. (1972). Propagation by roots. *Horticulture,* **50(5)**: 57-61.

Ramadasan, A. (1994). Vegetative propagation of spices. In: Advances in Horticulture, Vol. 10-Planatation and spice crops Part 2. Eds: K.L Chadha and P. Rethinam. Malhotra Publishing House, New Delhi.

Randhawa, G.S. and Mukhopadhyay, A. (1986). Floriculture in India. Allied Publishers Pvt. Limited, New Delhi.

Rao, P.S., Heble, M.R. and Chadha, M.S. (1980). Plant tissue culture, genetic manipulations and somatic technology of plant cells. BARC, Bombay.

Reddy, K.K. (1992). Vegetative propagation and biotechnology for tree improvement, Natraj Publisher, Dehradun, India.

Reddy, Y.T.N., Jankiram, T. and Reddy, D.S. (2001). Scientific nursery management. The House of Sapran (Media), Bangalore.

Reiley, H. E. and Caroll, L.S. Jr. (1981). Introductory Horticulture. 4th edition. Delmar Publishing Inc., Albany, New York.

Reinert, J. and Bajaj, Y.P.S. (1977). Application and Fundamental aspects of plant cell, tissue and organ culture. Spring-Verlag, Berlin, Germany.

Reinert, J. and Bajaj, Y.P.S. (1977). Plant cell, tissue and organ culture, Springer-Verlag, Berlin, London.

Rom, R.C. and Carlson, R.F. (1987). Rootstocks for fruit crops. John Wiley, New York.

Sadhu, M.K. (1986). Plant Propagation. New Age International Publishers, New Delhi.

Sen, S.K. and Giles, K.L. (1982). Plant cell culture in improvement, Plenum press, New York.

Sharma, R.R. (2002). Propagation of horticultural crops. Kalyani publishers, Ludhiana, India.

Sharma, R.R. (2004). Propagation of horticultural plants. Farm bulletin, Directorate of Extension, Ministary of Agricultural, Govt. of India.

Sharma, R.R. and Srivastav, M. (2004). Plant propagation and nursery management. Intl. Book Distributing Co., Lucknow, India.

Sharp, W.R., Larsen, P.O., Paddock, E.F. and Raghvan, V. (1979). Plant cell and tissue culture: Principles and applications. Ohio State University Press, Columbus.

Srivastava, S.S. (1991). Handbook of Horticultural Science, 3rd edition, Central Book House, Sadar Bazar, raipur-492 001.

Street, H. E. (1977). Plant tissue and cell culture, 2nd ed., University of Calfornia Press, Berkeley, U.S.A.

Taylorson, R.B. and Hendricks, S.B. (1977). Dormancy in seeds. *Annual Review Plant Physiol.,* 28: 231-254.

Tewari, J. P. and Mishra, K.K. (2003). Nursery Production and management of sub-tropical fruits, Training manual, Department of Horticulture, G.B.P.U.A.&T., Pantnagar, Uttaranchal.

Thorpe, T.A. (1981). Plant tissue culture : Methods and applications in agriculture. Academic press, New York.

Tilney-Basset, R.A. (1963). The structure of periclinal chimeras. *Heredity,* **18**: 265-285.

Tukey, H.B. (1964). Dwarfed fruit trees. Mac-Millan, New York.

Vasil, I. K. (1980). Perspectives in plant cell and tissue culture. *Intl. Rev. Cytol. Suppl.,* 11B, Academic press, New York.

Villiers, T.A. 91972). Seed dormancy. *In:* Seed biology, Vol. 2, (T.T. Kozlowski, Ed), Academic press, New York, pp.220-282.

Welch, H. J. (1970). Mist propagation and automatic watering, Faber and Faber, London.

Model Question Paper

1. Best planting material of pineapple
   a. slips b. crown
   c. stump bits d. runners
2. Propagation of wood apple is by
   a. root cutting b. stem cutting
   c. leaf cutting d. inarching
3. Tip layering is a successful method of propagation in
   a. mango b. litchi
   c. lemon d. black raspberry
4. Filbert can be propagated through
   a. layering b. stone grafting
   c. a & b d. cutting
5. Which of the following is used for the repair of a damaged stem
   a. bridge grafting b. layering
   c. shoot tip grafting d. microbudding
6. For production of virus-free planting material, which method is preferred?
   a. shoot tip grafting b. hard wood cutting
   c. gootee d. stooling
7. Air layering is most commonly used in
   a. banana b. mango
   c. *ber* d. litchi
8. Leaf bud cutting is practiced in
   a. apple b. *Pilu*
   c. walnut d. blackberry

9. Polyembryony is found in mango cultivar
   a. Dashehari   b. Langra
   c. Neelum   d. Kurrukkan
10. Stratification is associated with
   a. vegetative propagation   b. seed dormancy
   c. bud dormancy   d. none
11. How many layers of integuments are found in angiospermic plants?
   a. one   b. two
   c. three   d. four
12. Which kind of dormancy has been reported in *ber*
   a. mechnical dormancy   b. physiological dormancy
   c. morpho-physiological   d. deep dormancy
13. Which hormone promotes dormancy in seeds?
   a. ABA   b. GA
   c. Cytokinin   d. Auxin
14. Which hormone is responsible for the release of seed dormancy?
   a. GA   b. Cycocel
   c. ABA   d. AMO-1618
15. Which of the following is not used to overcome seed dormancy?
   a. stratification   b. scarification
   c. seed priming   d. tetrazolium test
16. Which of the following is used to estimate seed viability?
   a. Fluid drilling   b. Conductivity test
   c. Tetrazolium test   d. b & c
17. Radicle protrusion is the indication of
   a. seed dormancy   b. seed germination
   c. seed development   d. none
18. Epigeal germination is found in
   a. apple   b. mango
   c. custard apple   d. tamarind
19. Which quality of light favours germination
   a. green   b. far red
   c. red   d. blue

20. Select orthodox seed from the following
    a. mango b. mangosteen
    c. citrus d. papaya
21. Excised embryo test is used for
    a. breaking seed dormancy b. improving seed germination
    c. seed viability estimation d. all of the above
22. Mango is propagated through which method in *Konkan* region
    a. side veneer grafting b. inarching
    c. micropropagation d. epicotyl grafting
23. Propagation by apomictic seedlings is possible in
    a. apple b. guava
    c. ananas d. avocado
24. Apospory is found in
    a. *Rubus* b. guava
    c. pomegranate d. walnut
25. Bizzaria is an example of
    a. graft chimera b. transpons
    c. somaclonal variation d. topophysis
26. Topophysis is noticed in
    a. coffee b. tea
    c. rubber d. oil palm
27. Indicator plants are used for the detection of
    a. fungi b. bacteria
    c. viruses d. prions
28. Rhizocaline concept is associated with
    a. rooting of cutting b. seed germination
    d. graft healing d. seed dormancy
29. Which of the following does not promote rooting in cuttings
    a. basal wounding b. KNO_3
    c. girdling d. IBA
30. Which of the following is not a method of propagation by cutting
    a. eye cutting b. piping
    c. Irishman's cuttings d. Texas inlay

31. Four flap grafting is also known as
    a. banana grafting — b. awl grafting
    c. nurse-seed grafting — d. STG
32. Clonal rootstocks in apple are mostly propagated by
    a. inarching — b. seed
    c. budding — d. stooling
33. A clone can not be maintained by
    a. sexual propagation — b. cutting
    c. layering — d. micropropagation
34. Ultra dwarfing rootstock of apple is
    a. M27 — b. M111
    c. MM27 — d. M9
35. Cv. Anupam is used as an interstock in
    a. banana — b. mango
    c. guava — d. grape
36. Muscadine grape is propgated by
    a. cutting — b. serpentine layering
    c. grafting — d. patch budding
37. Which one is most stable form of chimera?
    a. periclinal chimera — b. mericlinal chimera
    c. sectorial chimera — d. all
38. Temerate crops are mostly propagated by
    a. tongue grafting — b. layering
    c. budding — d. seeds
39. Phalsa is commonly propagated by
    a. seeds — b. cuttings
    c. saddle grafting — d. ring budding
40. Adventitious embryony has been reported in
    a. almond — b. jamun
    c. pistachio — d. apricot
41. Hormone responsible for rooting of cutting
    a. Auxin — b. GA
    c. Kinins — d. Dormin

42. Which is known as rooting hormone

a. IBA b. IAA

c. NAA d. Phosphon

43. Which of the following is not a form of dormancy?

a. primary dormancy b. secondary dormancy

c. double dormancy d. topophysis

44. Hammer cutting is a method of a propagation for

a. Goose berry b. *Annona*

c. Malphigia d. Pecan nut

45. Scarification does not involve

a. rubbing b. sulphuric acid

c. hot water d. GA

46. Germination can not be promoted by

a. Charate b. KNO_3

c. PEG d. ABA

47. Stratification is done in

a. apple b. citrus

c. mangosteen d. fig

48. Vegetative propagation is not commonly practiced in

a. *phalsa* b. papaya

c. *aonla* d. a & b

49. Clonal variations of Anab-e-Shahi resulted in selection of

a. Tas-e-Ganesh b. Sonaka

c. Manik Chaman d. High Gate

50. Thornless blackberries are example of

a. chimera b. topophysis

c. transposons d. epigenetic variations

Answers to multiple choice questions

1	a	26	a
2	a	27	c
3	d	28	a
4	a	29	b
5	a	30	d
6	a	31	a
7	d	32	d
8	d	33	a
9	d	34	c
10	b	35	b
11	b	36	b
12	a	37	a
13	a	38	a
14	a	39	a
15	d	40	b
16	d	41	a
17	b	42	a
18	d	43	d
19	c	44	a
20	d	45	d
21	c	46	d
22	d	47	a
23	a	48	a
24	a	49	a
25	a	50	a

Match the following

Propagation methods			
1	Banana	A	Seed
2	Mango	B	Veneer grafting
3	Guava	C	Epicotyl grafting
4	Grape	D	Hard wood cutting
5	Strawberry	E	Runners
6	Cashew nut	F	Sword suckers
7	Almond	G	T-budding
8	Date palm	H	Suckers
9	Jack fruit	I	Inarching
10	Cape gooseberry	J	Stooling

1	2	3	4	5	6	7	8	9	10
F	B	J	D	E	C	G	H	I	A

Rootstocks/interstock			
1	Apple	A	*Psidium friedrichsthalianum*
2	Pear	B	*Khirni*
3	Mango	C	GF-577
4	Guava	D	Flying Dragon
5	Almond	E	M-9
6	*Ber*	F	*Zizyphus rotundifolia*
7	Citrus	G	Old Hume
8	Sapota	H	Ripario
9	Grape	I	Esmeralda and Irwin
10	Walnut	J	Paradox

1	2	3	4	5	6	7	8	9	10
E	G	I	A	C	F	D	B	H	J

Seed dormancy			
1	Plum	A	Physical dormancy
2	Seed coat dormancy	B	Mechanical dormancy
3	Pits of stone fruits	C	GA
4	Hard seed coat + dormant embryo	D	Moist chilling
5	Dormancy release	E	ABA
6	Dormancy induction	F	Double dormancy
7	Seed coat dormancy	G	Chemical dormancy
8	Pre-drying	H	77-100 °C
9	Hot water scarification	I	37-40 °C
10	Chemical inhibitors	J	Impermeable seed coat

1	2	3	4	5	6	7	8	9	10
D	A	B	F	C	E	J	I	H	G

Seed propagation			
1	Tetrazolium chloride	A	Reduced seed life
2	Seed viability	B	Plum
3	Orthodox seed	C	Formazone
4	Recalcitrant	D	Germination test
5	High temperature and high humidity	E	Sodium alginate
6	Seed vigour	F	Litchi
7	Fluid drilling of seed	G	SGRT
8	Brassinosteroid	H	2n
9	Sporophytic plant	I	ABA
10	Vivipary	J	Bio regulator

1	2	3	4	5	6	7	8	9	10
C	D	B	F	A	G	E	J	H	I

Vegetative propagation			
1	Co-factors	A	Basal cutting
2	IAA	B	Rooting of cutting
3	Improved rooting	C	Old Hume
4	Irish man's cutting	D	Labile hormone
5	Polarity	E	Juvenility
6	Localized incompatibility	F	Root grafting
7	Attached method	G	Proximal and distal ends
8	Bench grafting	H	Nurse root grafting
9	Temporary rootstock	I	Green grafting
10	*Phylloxera*	J	Inarching

1	2	3	4	5	6	7	8	9	10
B	D	E	A	G	C	J	F	H	I

Tissue culture			
1	Asepsis	A	Gibberellins
2	Hardening or acclimatization	B	*Agrobacterium rhizogenes*
3	Shoot proliferation	C	*In vitro* propagation
4	Shoot elongation	D	Cytokinin
5	Rooting	E	Citrus
6	Meristem culture	F	Stage IV of micropropagation
7	Embryo culture	G	Colchicines
8	Cell culture	H	Indirect organogenesis
9	Shoot/root development from callus	I	Seedless grape
10	Anther culture	J	Virus free plants

1	2	3	4	5	6	7	8	9	10
C	F	D	A	B	J	I	E	H	G

Propagation methods			
1	Truncheons	A	Mango
2	Torn cutting	B	*Ber*
3	Root cutting	C	Pecan nut
4	Leaf bud cutting	D	Apple rootstocks
5	Air layering	E	Bread fruit
6	Tip layering	F	Lemon
7	Patch budding	G	Raspberry
8	Trench layering	H	Quince (dwarf)
9	Ring budding	I	Fig
10	Top working	J	Macadamia

1	2	3	4	5	6	7	8	9	10
I	H	E	F	J	G	C	D	B	A

Seed propagation									
1	Mangosteen				A	Germination enhancement			
2	Vellaikolumban				B	X ray			
3	Psorosis				C	Parthenogenesis			
4	Haploid				D	Adventitious embryony			
5	Rest phase				E	Longer storability			
6	Thiourea				F	AAT			
7	Cryoprotectant				G	Seed borne			
8	Orthodox				H	Dormancy			
9	Seed vigour				I	Gametophyte			
10	Seed viability				J	DMSO			
1	2	3	4	5	6	7	8	9	10
C	D	G	I	H	A	J	E	F	B

Techniques and concerned scientists									
1	Bottom heat for mango propagation				A	Jumping genes			
2	Soft wood grafting				B	Auxin			
3	Veneer grafting in mango				C	Etiolation			
4	Hiltner				D	Lynch; Mukerjee			
5	Barbara McClintock				E	Amin and Co-workers			
6	Guha and Maheshwari				F	Reddy and Majumdar			
7	Epicotyl grafting in mango				G	Traub and Auchter; Bhan			
8	Sachs				H	Rooting co-factor			
9	Went				I	Brick gravel test			
10	Bouillenne				J	Anther culture			
1	2	3	4	5	6	7	8	9	10
F	E	D	I	A	J	G	C	B	H

In vitro propagation									
1	Biotization				A	Weaning			
2	$HgCl_2$				B	Preparation of mother plant			
3	Clone				C	Unorganized parenchymatous cell			
4	Culture medium				D	Sterilization			
5	Synthetic seeds				E	Phenolic exudation in medium			
6	Woody perennials				F	True-to-type			
7	Agar-agar				G	Somatic embryo			
8	Callus				H	MS medium			
9	Stage 0				I	Beneficial microbes			
10	Hardening				J	Gelling agent			
1	2	3	4	5	6	7	8	9	10
I	D	F	H	G	E	J	C	B	A

True or false

1. Seed propagation is most prevalent method of propagation in nature.
2. In modern fruit production system, seed propagation is most preferred method for raising rootstocks.
3. Apomixis has been reported in pomegranate.
4. The best planting material for banana is water suckers.
5. Kiwifruit is mostly propagated by hard wood cuttings during winters.
6. Tahiti lime produces nucellar seedlings.
7. Plumule is a rudimentary shoot.
8. The ratio of dry seeds to acid should be 1:2 for the acid scarification.
9. To overcome the seed dormancy, soaked seeds are prechilled at 5-10 ^{0}C.
10. Both, epigeal and hypogeal germination can be found in the same crop.
11. Cold test is used for measuring the seed viability.
12. Fluorescein diacetate (FDA) test is used for measuring seed vigour.
13. Hiltner test is performed to screen out the *Fusarium* infected weak seeds in cereal crops.
14. Seeds of tropical and sub-tropical fruits are sown during monsoon (June-July) or in the beginning of spring (Feb-March).
15. Sowing of temperate fruits seeds should be done in Feb.-March.
16. Seeds of loquat should be sown immediately after extraction.
17. Seeds of citrus can be stored for 5 years at ambient storage.
18. *In situ* vegetative propagation is practiced to reduce mortality and improve survival of graftage/budded plants.
19. Sexual propagation always results in production of true-to-type plants.
20. Vegetatively propagated plants are less vigorous and thus their management is easy and less costly than seedling plants.
21. The best way to perpetuate a clone' is seed propagation.
22. Virus indexing technique is possible through cuttings.
23. Inarching is followed to repair the damaged root systems of plants.
24. Vegetatively propagated plants, particularly budded and grafted are short lived in comparision to seed propagated plants.
25. A new variety can be developed by vegetative means of propagation.
26. In diplospory, reductional meiosis is replaced by a restitutional meiosis or by mitosis.
27. *Malus toringoides* and *M. sikkimensis, M. huphensis* are non-apomictic rootstocks of apples.

28. Dashehari-51 is a clonal selection from cv. Baneshan.
29. Transposable elements may bring about genetic variation in a clone.
30. Etiolation helps in amelioration of seed dormancy.
31. Perlite is used to measure seed viability.
32. Precuring is done for the preparation of layers.
33. Grafting is a very successful method of propagation in monocots.
34. Matching up of scion and stock cambium is necessary for successful graft union.
35. Delayed incompatibility is noticed in black lines of walnut.
36. Wedge grafting is an approach method of grafting.
37. Budding is a quick and efficient method of propagation as compared to grafting.
38. Fall budding is done in late summer rather than fall.
39. In T budding, 'T' is the shape of the bud and 'shield is the shape of cut given on the rootstock.
40. GA_3 is commercially used for the production of layered plants.
41. Etiolation is done to improve graft healing in citrus.
42. Haberlandt is known as the father of tissue culture technique, who for the first time coined the term totipotency.
43. Pumice is of volcanic origin and mined in to commercial grade as a potting medium.
44. Hormonal powders are used for rooting of cuttings.
45. 'Prolonged soaking method' is used for *in vitro* multiplication of fruit crops.
46. Pinching of buds is done to encourage budding success in apple.
47. Anther culture is followed to produce polyploidy plants in fruit crops.
48. Lanoline paste is used in tissue culture of banana and papaya.
49. Turanj produces apomictic seedlings.
50. Mycorrhization/biotization helps in improving plant survival during hardening of tissue cultured plants.

Answers (True or false)

1.	T	2.	F	3.	F	4.	F	5.	T	6.	T	7.	T
8.	T	9.	T	10.	F	11.	F	12.	F	13.	T	14.	T
15.	T	16.	T	17.	F	18.	T	19.	F	20.	T	21.	F
22.	F	23.	T	24.	T	25.	F	26.	T	27.	F	28.	F
29.	T	30.	F	31.	F	32.	F	33.	F	34.	T	35.	T
36.	F	37.	T	38.	T	39.	F	40.	F	41.	F	42.	T
43.	T	44.	T	45.	F	46.	F	47.	F	48.	F	49.	T
50.	T												

Write short notes on:

1. Air layering in litchi
2. Veneer grafting in mango
3. Epicotyl grafting
4. Soft wood grafting for *in situ* orchard establishment
5. Bottom heat technique
6. Seed dormancy
7. Epicotyl dormancy
8. Double dormancy
9. Acid scarification
10. Stratification
11. Seed priming
12. Cryopreservation
13. Excised embryo test
14. Tetrazolium test
15. Apomixis
16. Clone
17. Chimera
18. Rhizocaline concept
19. Topophysis
20. Bizzaria Orange
21. Polarity in grafting
22. Micrografting
23. Synseed
24. Stages of micropropagation
25. Stionic effect

Distinguish between following with suitable examples;

1. Sexual vs asexual propagation
2. Epigeous vs hypogeous germination
3. Primary vs secondary dormancy
4. Nucellar vs zygotic seedlings
5. Orthodox vs recalcitrant seeds
6. Seed vigour vs seed viability
7. Aposory vs diplospory
8. Periclinal chimera vs mericlinal chimera
9. Depotting vs repotting
10. Germination test vs tetrazolium test

Plate 1. Side grafting in mango

Plate 2. Top working in apple.

Plate 3. **Stool layering in apple clonal rootstocks.**

Plate 4. ***In vitro*** **shoot culture of kiwifruit.**

Plate 5. **Hardening of micropropagated grape under protected conditions.**

Plate 6. ***In vitro* somatic embryo development in mango.**

Plate 7. **A well managed apple nursery.**

Plate 8. **Polyethylene tunnel cultivation.**

Subject Index

T

V

W

X